GRAVITY FROM A

NEW ANGLE

Karunakar Marasakatla

ISBN-13: 978-0-98197-687-7
ISBN-10: 0-98197-687-5

Version 1.2
First paperback publication: August, 2009

Published by Karunakar Marasakatla
www.kmarasakatla.com

To my family

Soumya, Sumeda and Sunitha

for their continuous and unconditional support

all along the course of my work.

- Karunakar Marasakatla

Gravity From A New Angle

Table of Contents

Introduction

Gravity appears like a simple phenomenon because we experience it in every movement of our lives. It can displace an object from one place to another just like any other form of energy, but still we wouldn't agree the gravity as a source of energy. In reality, there is not much difference between other forces in nature and the gravity. Even though all forms of energy and gravity appear to be similar in all the functionality, theoretically they are poles apart. Gravity remained independent from all other known forces.

Because it was not clear how gravity expenses energy when it does work, a different nature and principles like curvature in space were attributed to it. Because of this unique and strange nature, it became difficult to merge it with other entities. In fact, all other forces are different forms of gravity itself. A simple misunderstanding of couple of basic principles in physics caused the science to be disjointed. This resulted in the formation of many branches in science without any common base between them.

The concept of force, definition of mass, inverse square law of gravity and shell theorem on which the entire physics is dependent on are all flawed. Because of the flaws in these concepts, gravity was misunderstood and was described as a weak force among the known four forces in the nature. Whenever another force was observed with a different nature, it was given names like the strong nuclear force, weak nuclear force, surface tension, capillary action, etc. and even went to an extent to define a new fifth force when an attraction was observed between two floating objects.

When these flaws in basic principles are removed, a clear picture of not only the physics but the entire science emerges because the physics was considered as the basis for all other sciences.

This book is divided into three parts. Part-I deals with the present understanding of the basic principles of physics and the reasons why we need to change them. It describes the current understanding of the gravity, Newtonian principle of gravity, General Relativity and the recent developments like MOND (Modified Newtonian Dynamics) and explores the flaws in these principles.

Part-II of the book deals with what is the change being proposed in this new theory. It defines the new principle of gravity and also explores the nature and functionality of the gravity, like where it is being stored and how it is expensed. In short, the new theory proposes that the gravity of an object increases as it compresses or appears smaller from another object. Gravity between two objects depends upon the "Angle" of projection of each object on the other. Basically, the definition of mass blindfolded the science in understanding the true nature of gravity.

Part-III of the book deals with how the new theory is applied to the nature. It concentrates on explaining the observed phenomena using the new theory. It also predicts other possibilities where we can apply and test this new theory in the future.

A simplest explanation for the pioneer anomaly is that the central object which exerts the gravity appears smaller at those distances, ultimately increasing the gravity on deep space objects on the outskirts of solar system.

As per the content level of this book, it was written in a simple language to make it understandable even with a basic knowledge of science.

References are mentioned in the square brackets, ex. [1]. Some of these references are general in nature and not necessarily a reference to the original work. Attempt was made to include references to all the major works referred in the text. If I had missed any references, they will be made available in the future revisions.

Finally, I like to thank my daughter, Soumya for valuable suggestions and spending part her summer vacation in editing the text.

Karunakar Marasakatla

Part I: Flaws in the Basic Principles of Physics

The prevailing theories in physics appear to have flaws in multiple fronts. The chaos in the field of gravity is the collective result of these flaws.

Modern history of gravity started from the works of Isaac Newton. Newton described gravity as an aspect of classical mechanics. Certainly, gravity appears to be part of mechanics because it does work instead of an imaginary concept of curvature of space around a massive body. Therefore, it is worth exploring the gravity from the perspective in which Newton observed the phenomenon.

This section explores the validity of the basic principles of mechanics such as the force and the resultant force, definition of mass and inverse square law of gravity. Then the prevailing theories of gravity were examined at the end of this part.

Chapter 1:

Why we need to change our beliefs?

The field of gravity is ripe with anomalies. No other fundamental force is so widely contested than this less understood force. There were many attempts at describing gravity and the principles governing it from the period of Isaac Newton to the present time. Each attempt either improved the understanding of gravity or complicated it further, but never cleared the cloud over it.

Today, gravity became an isolated force among all other fundamental forces. It is widely believed that the nature of all the fundamental forces should be able to be captured in a single principle. A unified theory of all the forces will evolve only when gravity is described in terms of all other forces or when all other forces are described in terms of gravity.

We think that there is no need to change the present principles of gravity because it successfully describes the celestial movement at the solar system scale. Certainly these principles were unable to describe the celestial dynamics beyond the solar system. Other stars in the galaxy are revolving around the galactic center beyond the speed predicted by the current principles. At that speed, stars will fly away from the galaxy because there is not enough matter at the galactic center or inside the orbit to bind these stars gravitationally to the galactic center.

In the absence of any satisfactory alternative theory to the prevailing standard theory, the present principles invented the dark matter [1, 2, 3] to describe the observed anomalous galactic movements. In spite of continuous search for decades, the particles which compose the dark matter were never found. To describe the observed phenomenon, the characteristics of the dark matter particles were continuously altered. Finally, the dark matter particles took the shape as Weakly Interacting Massive Particles (WIMPs). The distinction of the dark matter from all other known particles should also be in such a way that it should enable the proper movement of the stars around the galactic center without interacting with the visible matter. To make this possible, the dark matter was placed around the galaxy instead of within the galaxy. In reality, even with this placement of the dark matter, it is not possible to make the stars revolve around the galactic center.

The general perception about the dark matter was aptly summed up in a letter to a magazine [4] and compared it to the ether, the medium presumed to be spread all over the universe, which turned out to be non-existent.

The dark matter is similar to the dark planet that hit the proto-earth to form the moon. The speed and direction of the dark planet has to be in the required range and also the size and density of that planet has to be sufficient to give enough angular momentum to the early earth. More over the planet should be ready to go over its defined path towards the earth at the right time when the earth just finished the differentiation. We should be lucky enough to see these kinds of chance occurrences in millions of planetary systems.

So, whether we find the dark matter or not, the problems and anomalies still continue to exist for this amazing force we see everyday.

Pioneer anomaly, the unknown force pulling the spacecraft towards the sun, is another unknown frontier in the field of gravity. Local gravity anomalies or gravity hills around the world which were conveniently described as optical illusions, local gravity variations during the eclipse, anomalous pull of the earth on a spacecraft beyond the known gravitational pull, uncertainty in the value of the gravitational constant, G are some of the issues which require a satisfactory explanation in any theory that explains the gravity.

It is hard to believe that a strong force at micro level, like the strong nuclear force, disappears into thin air at macro level. The principles for the quantum world wouldn't work at the cosmic scale and the principles of gravity wouldn't work at the atomic level. It is widely believed that if matter is made of same materials then the principles describing it should work at both micro and macro levels. According to the present theories, these principles are mutually exclusive.

It is simply a self deception if we think that we really understand the nature of gravity even after all of these prevailing anomalies, including the isolation of gravity from all other fundamental forces.

The challenge ahead for any new principle of gravity is that it has to explain all the observed phenomena. It should also be able to resolve all the anomalies and predict new possibilities to explore further in the future and should unify the gravity with the existing fundamental forces.

The task ahead for the new theory appears to be an unattainable goal unless there is a fundamental mistake in our understanding of gravity. And it appears that we are doing the same mistake. Further in this book, the flaws in the existing theories are explored and a new principle is formulated to describe the nature of gravity along with new predictions to explore it further.

Chapter 2:

What is Inertia?

We take most of the basic principles in physics for granted. We never think of a flaw in these principles because these principles were being used for a long period of time. As we derive new principles and theories based on these basic principles, a large set of principles accumulates based on this small set of basic principles. Changing the basic principles will have an enormous effect on our entire understanding of science. To maintain the status quo, even after a conflicting discovery was made, the new discovery was accommodated into the existing theories with convenient extensions to the existing theories or termed the new discovery as an anomaly and left it in the cold. As these extensions and the anomalies grow in number, it will not be possible to explain all these theories in a coherent manner and eventually the entire system wouldn't make any sense. The present state of science is in similar situation. The reason for the present situation appears to be misunderstanding of multiple basic concepts in science. One of these misunderstood basic concepts is the inertia.

Inertia was described as the natural tendency of a resting object to resist a change in location. This can be best observed when we try to push a big resting object like a big round stone. Initially, it will be very hard to push the stone, but when it starts moving, it will be easy to push it further on a level field with less effort.

A resting stone has more inertia than the moving stone. Inertia is the fundamental aspect of the gravity but still there is no satisfactory explanation for why it is more with a resting object and less when it starts moving.

A stone is at rest on the surface of the earth because earth's gravitational force is continuously pulling the stone towards its center. The gravitational force is applied on the stone for as long as it is on the surface of the earth. Because gravity does some work on the surface of the earth, let's assume the gravity as a form of energy for the time being.

If the gravity is a form of energy, then a stone on its surface continuously consumes the energy as long as it is on the surface of the earth. Let's assume that it takes X Joules of energy per second to keep the stone on the surface of the earth. When we try to push the stone, the energy applied on it in any given second should be greater than the earth's gravitational energy being applied on the object in that particular second. Let's assume the energy applied on the object is Y Joules in each second. The forces acting on the stone were described in the Fig. 1. Instead of calling the entities in the figure as forces, we call them as energy consumed and give them a time period as well as a direction.

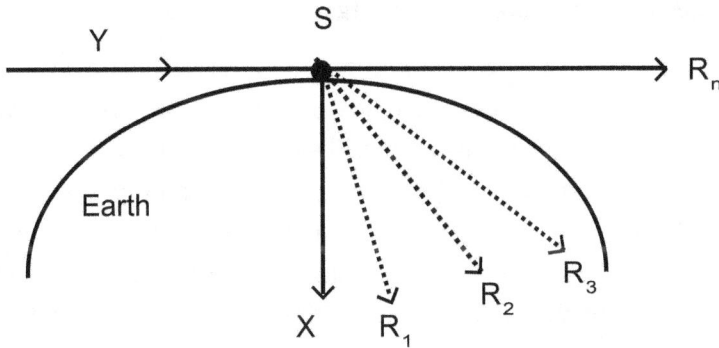

Figure 1: Inertia on an object

S is the stone being displaced on the surface of the earth. Let's assume that there is no friction on the surface where the object is being displaced. X is the gravitational energy applied on the object per second. Y is the external energy applied or transferred to the object per second. R is the net or resultant energy gained by the object in that particular second.

As long as the direction of the net energy gained by the stone points inside the earth (R_1, R_2, R_3), the object wouldn't move on the surface. All the energy gained by the stone will be used in compressing the earth beneath it. It means, as we push the object with less amount of energy than the gravitational energy holding the stone in each second, the object wouldn't move at all. This is what we observed as the Inertia. As soon as the external energy in any given second is enough to make the direction of the resultant force R, parallel to the surface of the earth as in R_n, the object starts moving and gains the kinetic energy equal to the R_n in that second.

The net energy gained by the object after the first second will be the difference between the resultant force and the loss of energy due to the friction for the distance it travelled. Now if we keep applying the same amount of energy Y in each second on the object, it will continuously accelerate and gains more kinetic energy.

For an object with a net kinetic energy of K Joules, it requires less external energy to displace the object because now there are three energy entities working on the object. They are the self kinetic energy of the object, gravitational energy and the external input energy. In this scenario, to keep the resultant force parallel to the surface of the earth, a less amount of external energy is required than to displace a resting object. This is why a moving object has less resistance or inertia compared to when the object is at rest.

To keep the initial energy used to overcome the gravity to a minimum, the object needs to start with the highest speed in the first second of its journey, like the flying golf ball when it was hit very hard in the beginning. Once the golf ball has started moving, we are not at all applying any energy to that object. It means, the object has started with highest speed. It also means that the gravity didn't hold it for long, so less energy was consumed in overcoming the earth's gravity in the first second. If an object picks up the speed after a long time, like a heavy truck gaining momentum after traveling for few minutes, lot of energy will be used in overcoming the effect of earth's gravity in that long period of time.

Therefore, the distance travelled by the object and the energy used on the object has no relation. Energy used by a system is dependent on the force or energy applied in each second and the period the energy was applied on the object.

Relation between Inertia, Speed and Distance

Inertia is nothing but the continuous grip of gravity on an object. If it takes more and more time to displace an object, gravity will work for that amount of time and we feel more affect of inertia. If the object moves very quickly, the grip of gravity on the object is for a short period of time, therefore less difficulty or less energy will be consumed in moving the object.

As shown in Fig 2, object A and object B are continuously accelerating towards the point Y from point X, a distance of 1 km. Object A took ten minutes to travel the distance of 1 km and had a speed of 10 km per hour at the end of the distance. Another object, B same as the object A, travelled the same distance of 1 km in less than ten minutes. Let's assume the time taken by the object B as 5 minutes.

Object A has to overcome the inertia, the affect of gravity, for a period of ten minutes whereas the object B has to overcome the affect of gravity for only 5 minutes to travel from point X to point Y. In this case, object A dispenses more energy to travel the distance of 1 km than object B to travel the same distance, moreover at the end of the distance, the object B will have more speed than the object A.

For an object to travel a given distance, the energy required will depend upon how long the object takes to travel that distance. More time it takes, more energy will be consumed. If we give all the energy consumed by the object A to travel from the point X to Y, to the object B at once in the beginning at point X, then the object B travels the distance quickly and at the same time gains more momentum at the end of the distance.

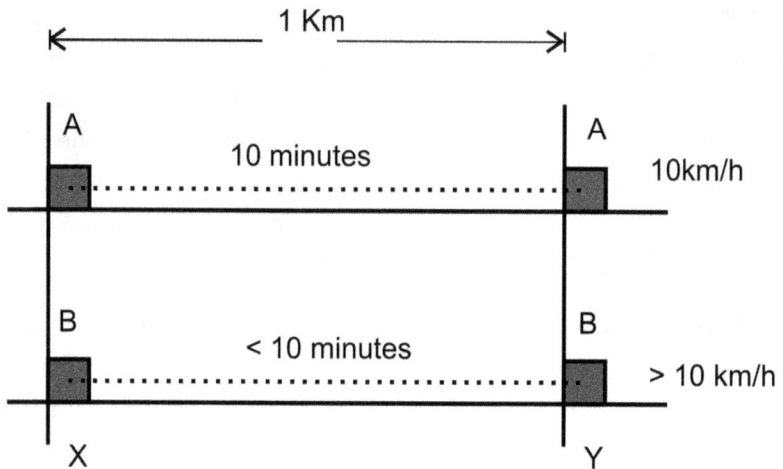

Figure 2: Inertia, speed and the distance

Recently the hyper milers, people who drive vehicles more distance for each gallon of gas (petrol) than normal, were in the news in the United States of America when the gas became expensive. It appears that these hyper milers have more awareness about the relation between the speed of the vehicle and the inertia, the grip of gravity. To drive for a maximum distance, these hyper milers avoid the road where there are STOP signs and traffic lights on their way by selecting alternate routes. It takes more energy to accelerate the vehicle to the earlier speed from a complete stop and wait at the STOP sign and also at the traffic light.

How an object comes to a halt?

As shown in Fig 3, K is the total kinetic energy of the object S and X is the gravitational energy on the object per second. The resultant force of these two forces will slowly point towards the center of the earth from the initial position of parallel to the surface of the earth. When the resultant force points inside the earth, the object keeps pushing the earth inside instead of moving forward. The

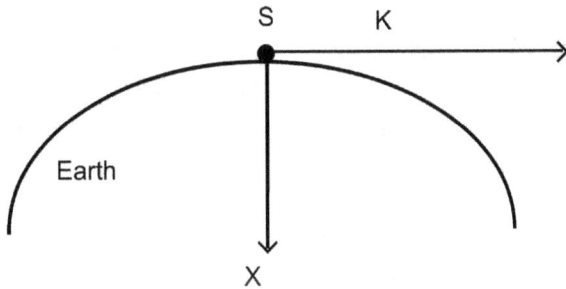

Figure 3: Object coming to a halt

objects kinetic energy will decrease as the object starts to push the earth straight down and compress the surface of the earth in each second. The distance travelled by the object will decrease with the decreased total kinetic energy of the object in each second. Slowly, when the kinetic energy is equal to zero and the resultant force points straight towards the center of the earth, then the object comes to a halt.

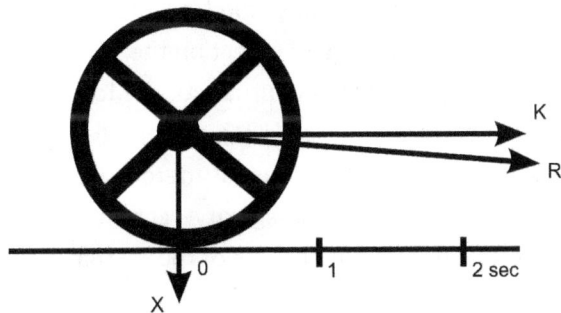

Figure 4: Forces on a moving wheel

How a wheel balances while moving?

A fast moving wheel always balances by itself. When it slows down to a point, it starts to loose its balance and eventually falls to the ground. The wheel and its applications are enormous but still we don't have any satisfactory explanation for the balancing of a moving wheel.

As shown in Fig 4, *X* is the total energy expended by the earth in a second to hold the wheel on to the surface. *K* is the total kinetic energy of the wheel when it started. *R* is the resultant force of these two forces at the end of the second. And again *R* will be the new kinetic energy of the object in the beginning of the next second.

As long as the resultant force is parallel to the ground, in other words when the object has more kinetic energy, it keeps moving forward with high speed. When there is no external input of energy into the wheel, the resultant force slowly points towards the earth.

A fast moving object has less grip of gravity in any given second. So the wheel retains its initial position because no other strong force than its kinetic energy is acting on it. When the wheel has less kinetic energy, the resultant force points towards the earth because the strength of gravity pulling down the wheel exceeds the kinetic energy of the wheel; as a result the wheel looses its balance and falls to the ground.

Inertia can be best described as the loss of energy in the earth in continuously holding an object on the surface of the earth. The effect of inertia is a time dependent function. More and more time gravity holds an object; more energy is required in displacing that object. The initial assumption of gravity as a mere force of a field and the later assumption as an imaginary curvature in space hindered the understanding of true nature of the inertia.

Chapter 3:

Force, Power, Work and Energy

Strong gravity is able to displace an object placed in its vicinity. It indicates that the gravity does some work. As work originates from the energy, we can consider the source of gravity as a source of energy. In this scenario, it is worth revisiting the concepts and theories relating to the energy and its consumption.

As discussed in the chapter on Inertia, energy consumed in a system is dependent on the force applied and the period of time the force was applied, not on the distance the object got displaced. An earlier work also argues that the work performed is dependent on the energy consumed in the system, not on the displacement of the object [5].

Energy content of a compressed or expanded objects of same amount of material are not same. A compressed object will

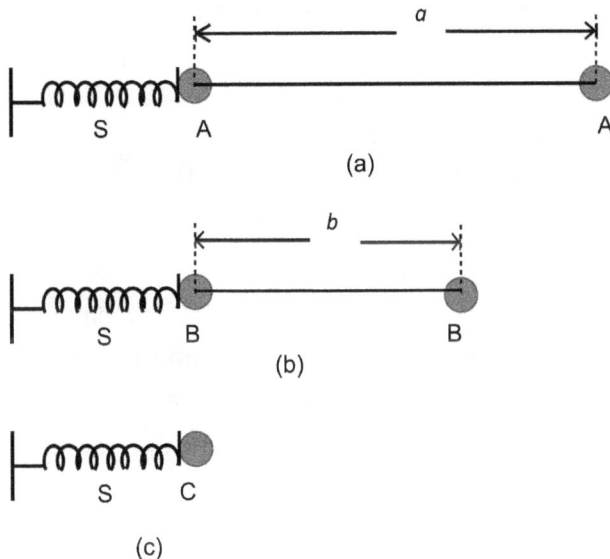

Figure 1: Energy and distance travelled

have more binding energy and an expanded object will have less binding energy. Energy level of the object not only depends on the displacement but also on the change in size and shape of the object. Energy consumed in a system or the amount of work done in a system should account for all these changes in the system.

To understand the scenario, let's push different objects with varying mass by a single compressed spring *S* as shown in the Fig 1.

Lets assume that the spring *S* stores *X* amount of energy when compressed. When the compressed spring is released, it hits the object *A* of 1 kg of mass as shown in Fig 1(a) and transfers the energy stored in it to that object. As a result, object *A* moves to a distance of *a* meters.

Let's take another scenario with two different objects *B* and *C* weighing 10 kg and 100 kg respectively as shown in Fig 1 (b) and (c).

In the case of object *B*, it travels a distance of *b* meters which is less than the distance travelled by object *A* with same amount of initial energy as the object *A*. In the case of object *C*, let's assume that the spring couldn't able to move the object at all. Object stays at the same point. In these three scenarios, the force applied on the target objects and the period the force applied as well as the energy transferred to these objects is same but the distance travelled by each of the individual objects is different. In the third case, the object didn't even move. Still, even in this third case, energy was transferred to the object.

If we take the standard equation for the work, *work = force * distance,* and apply that to the energy transferred in this system, the third scenario projects that there is no transfer of energy to the object because the target object didn't even travel any distance.

To push an object to certain distance, different forces will use different amount of energy. Due to the inertia caused by the gravity on the object, the system uses more energy when the force applied on the object is low, because it takes more time to push the object to a specific distance. When the force is high, it takes less energy to move the object for the same distance because it takes less amount of time, where it has to counter balance the earth's gravity on the object for a less amount of time.

Karunakar Marasakatla

After observing these scenarios, it is evident that the definition of Joule, the energy required to displace a one kilogram object to a distance of one meter is flawed. Along with the definition of Joule, the standard definition of work also appears to be flawed.

Joule should be defined as a fixed amount of energy and it should not depend on the distance. The definition of work also should not depend on the displacement of an object. The amount of work done should be defined as the total amount of energy consumed in the system. Energy transferred or the work done should be measured as the rate of energy transfer multiplied by the time applied. In current definitions, the rate of energy transferred was defined as power (Joules/second).

Work should be completely dependent on the power applied. In the present definitions, the power is derived depending upon the amount of work done in a system. There is no relation between the actual amount of energy consumed and the derived power. Power should be associated with the object generating energy not on change in the object on which the energy is applied. Different objects will react differently to the apply of same amount of energy or power. Therefore the correct representation of work in a system should be the total amount of energy consumed in the system.

If an object got compressed instead of displaced when energy was applied on the object, the effect of compressing the object should also be considered as work done. A compressed object gains more binding energy than the normal object.

What is the difference between power and force?

According to the current standards, the definitions of power and the force are as follows:

Power = rate of energy transfer per second.

= work / Time taken to do that work.

The power is measured in Joules per second or Watts.

*Force = mass * acceleration*

The force is measured in Newtons.

A moving object has speed, mass and total energy associated with that object. When there is a transfer of energy from this object to another object, power of the object comes into picture. It is hard to imagine the significance of force of the object to any of the physical characteristics of the object.

To clarify the significance of force in a system, let's first assume that it does exist in a physical system and for example is being represented as X Newtons of force in the swing of golfer's club. Lets also assume that if that club hits the ball, the rate of energy transfer from club to the ball as y Joules/second. If the golfer swings the club at the ball and misses it then there is no transfer of energy even though there is a force of X Newtons in the swing of a club. It means, unless there is a transfer of actual energy, there is no significance to the force in the system. Force has significance only when there is a transfer of energy. Force and power are intertwined. We even use these two terms synonymously in our day to day references to the strength in a system.

The rate of energy transfer is proportional to the force applied. When more force is applied, greater the power in the system and lower the power in the system means less force is applied. Neither of them has significance without the other.

Even the force meter, by which we measure the amount of force in a system, requires the expense of energy to measure the amount of force in a system. The rate of energy transfer determines the amount of force in the system. The more we expense energy in pulling the spring in the force meter, the more it measures the amount of force-Newtons.

To further explore the relation between force and power, lets consider a water tank with an outlet at the bottom and a stopper for that outlet. When the outlet is open, X amount of water will be released from the outlet per second. Assume that the water level of the tank is always the same. If we

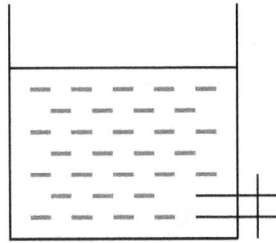

Figure 2: Water Tank

close the outlet, there is no release of water from the tank, even though there is a force at the bottom of the tank due to the water level. If we again open the outlet of the tank, then same amount of water will be released from the tank as earlier in each second.

The force of an entity, such as a water column, is something we can't see as an individual entity. It can only be felt with the energy the entity transfers to the other objects. Therefore we can represent the force in a system with power. Wherever the force is used in equations, it should be redefined as power, an energy transfer of Joules/second. We could get rid off the concept of force and its unit - Newton, altogether.

Work should be redefined as power multiplied by the time.

$$Work = Power * Time \qquad\qquad \text{--- Eq. 1.}$$

To understand in more detail about the energy consumed in a system, let's explore another scenario.

Let's consider the consumption of energy when an object travels from point A to point B with varying speed at different intervals as shown in Fig 3.

Object X is travelling at a constant speed higher than object Y all along the distance from point A to point B. Object Z travelled the distance from point A to point B from rest to rest with acceleration till the midpoint and later used its kinetic energy to travel the rest of the distance. It means, object Z consumed energy only till the mid point of the distance. Objects X and Y were continuously using the energy all along their travel from point A to point B. In this case object X takes less time

Gravity From A New Angle

to travel the distance than object Y and object Y takes less time to travel the same distance than the object Z.

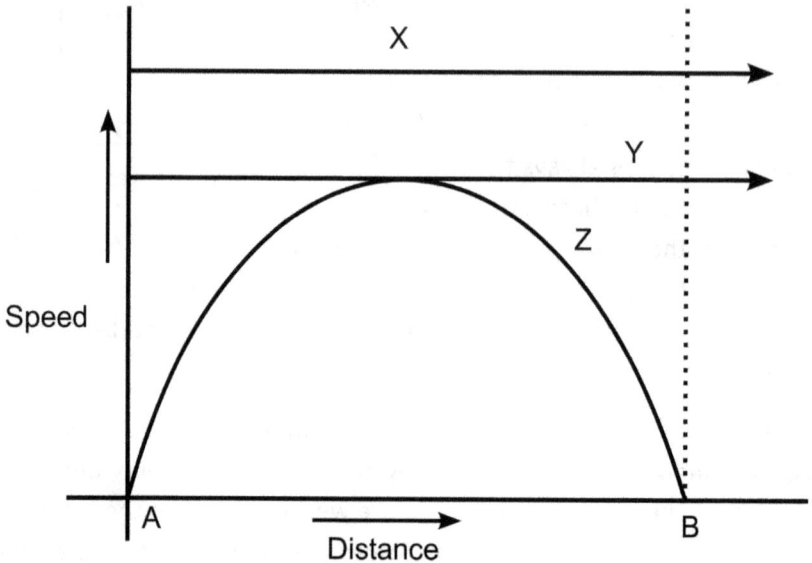

Figure 3: Work and Speed

As we have seen earlier, inertia on an object will increase when the speed of an object is lower and if the speed of the same object is higher then the inertia on that object will decrease. When inertia on an object is low, energy required to move that object will also be less. Object X uses less energy to travel the distance from point A to B when compared to Y. Object Y uses less energy to travel the same distance when compared to object Z.

Therefore in these three cases, objects X, Y and Z uses different amount of energy to cover the same distance from point A to point B. It is now evident from these scenarios that the energy consumed in the system doesn't depend upon the distance travelled.

Is energy should be considered as a standard unit?

Energy appears to be an independent entity. It is not dependent on the distance, mass or time. Therefore, we should consider defining the energy as one of the fundamental units.

Current definition of Joule depends upon the displacement of an object for a distance. As the distance based theories are flawed, we need to redefine the unit for the energy independent of these variable entities.

A Joule should be a quantity of energy which shouldn't depend on a mechanical system whether it is an expansion, contraction, displacement or any other activity.

Definition of Power

Power is the rate of transfer of energy from one object to another. The rate of transfer depends upon the area of contact between those two objects. In this case, the power can be best described using the parameters energy, time and the area of contact.

$$Power = Energy / (Time * AreaOfContact). \qquad \text{--- Eq. 2.}$$

$$Work = Power * AreaOfContact * Time. \qquad \text{--- Eq. 3.}$$

All the three concepts, force, work and power were gradually defined in the works of different scientists over a period of century. Newton defined the concept of force. Work, power and energy were not at all part of the Newtonian mechanics. Power was defined almost a century later by James Watt. Mechanical system is transfer of energy from one entity to another. Without the concept of energy and power, we can't describe a mechanical system. Most probably, Newton might have used the concept of force in similar to the present definition of power. Later, because the true nature of gravity was not understood, the overlap or the similarity between the definitions of force and power was not identified. The concept of work was also incorrectly defined based on the prevailing understanding of gravity at that time.

Gravity From A New Angle

Now on, the words Force and the Power will be used synonymously in the rest of the book. Wherever the word force is used, the underlying meaning of the word will be assumed as Power. It will also be same when the word is in combination with other words, like the resultant force; which will be in fact a resultant power.

Chapter 4:

Concept of Resultant Force

A resultant force is the combined effect of the multiple forces acting on a point. Unless the angle made at the point by the forces is equal to zero, the resultant force will never be equal to the net value of the forces. It will always be less than the combined forces when the angle is greater than zero. When the angle at the point of action, where the forces interact, is equal to zero then the resultant force will be equal to the sum of the forces.

As shown in Fig 1 (a), force F_1 expenses 4 Joules/sec and F_2 expenses 6 Joules/sec in displacing the object from point A to point C. Here, both the forces are acting at point O on the object at right angle. So the

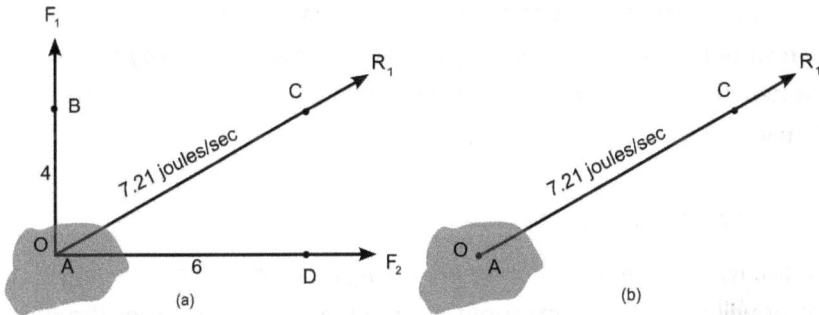

Figure 1: Difference between single and multiple forces at a point

combined expense of the energy in moving the object by both the forces is equal to 10 Joules/sec.

Let's assume that the object moves from point A to point C in a period of one second. According to the principles of statics, if we calculate the value for the resultant force of these two forces then the value of the resultant force, R_1 will be equal to 7.21 Joules/sec, less than the sum of the individual forces. It means if it would have been a single force, as the resultant force R_1, acting on the same object at point O then the energy expense would have been just 7.21 Joules/sec to displace the same object to same distance in one second as shown in Fig 1 (b).

In this case, it is evident that when multiple forces are acting at a point on an object, more energy will be consumed than when less number of forces acting on the same object to do same amount of work.

Current theories will proclaim that only 7.21 Joules/second was consumed in the system in both the cases, when F_1 and F_2 were acting on the point in Fig 1 (a) and a single force equal to the resultant force R_1 was acting on the point as in Fig 1 (b).

In reality, the amount of work done in Fig 1 (a) in moving the object from point A to point C is just 7.21 Joules/second in a period of one second. Actual energy expensed is 10 Joules/second. The difference between the energy expensed in the system and the energy consumed in displacing the object, a 2.79 Joules/second, is unaccounted for in the system. Where is this energy disappeared into? According to the same conventional theories, it is not possible to create or destroy the energy; we can only transfer it from one form of energy into another form. Then into which form the missing energy got transferred to?

Differential Force

Basically, when multiple forces are acting on a point in a wide angle, the point will be under tremendous stress regardless of whether the object is moving or stationary. The stress on object, if the forces are continuously applied, will split it apart. Therefore the energy unaccounted for in the Fig 1 (a) is the same energy that subjects the stress on the object. We name this new force as the "Differential Force", the difference between the sum of all forces and the resultant force. Differential force works in the line of original forces proportionate to their strengths.

Differential Force, D_f = Sum of the Forces – Resultant Force

$$D_f = T_f - R_f \qquad \text{--- Eq. 1.}$$

For the Fig. 1(a), the differential force is

$$D_f = (F_1 + F_2) - R_1$$

Therefore the net result of the forces acting in Fig 1 (a) and Fig 1 (b) is not the same, even though they both displaced the object to a same distance in the same amount of time in the same direction. When the net total of the forces is more than what is required to displace the object, the difference in the forces will destabilize the internal structure of the object being pulled. The forces will basically either split the object or crush it depending upon whether the forces acting away from the object or toward the object.

Conventional theories are basically concentrated only on the net result. If we pull an object by applying multiple forces on the object from multiple directions, then the object wouldn't even move from the resting point if all the forces are in equilibrium. According to the

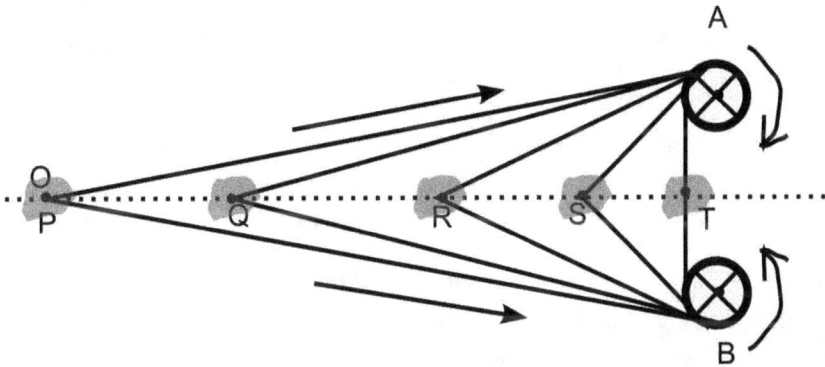

Figure 2: Resultant Force and the Differential Force

conventional theory, there is no work done in this scenario. But if we continuously apply the forces, the object will split apart at the point where the forces are being applied. Is this action of splitting the object into small pieces also considered as the work done? Answer for that question is yes because splitting the object also requires energy.

The relationship between all the forces acting on a point and the resultant force is illustrated in the Fig 2.

Two ropes are attached to the rim of two rotating wheels *A* and *B*. The other ends of the ropes were tied to an object at point *O* on the object. Initially, the object is at point *P*, away from the wheels and on the line going between them. Each rope is being pulled with equal force, meaning the energy sources connected to the ropes expend same amount of energy in a given period of time. Initially, the object being pulled will move faster towards the center point *T* between the two wheels. As the object approaches the point *T*, the angle made by the ropes on the object will increase and the speed of the object will decrease because the resultant force of the two forces will decrease. As the resultant force decreases, the differential force will increase. When the object is right in the center at the point *T*, the resultant force will be equal to zero therefore there wouldn't be any displacement in the object but still the forces will continue to dispense the energy on the object. In this case, the differential force will be equal to the sum of both the forces because the resultant force is equal to zero. The differential force will stretch the object and eventually will break it.

When a loosely coupled object, like a comet, approaches a huge planet like the Jupiter, the comet will split into pieces when it is close to the planet. Comet Shoemaker-Levy broke into pieces due to the differential force of the Jupiter, not because of the tidal forces.

A Roche limit, the radius around a planet within which an object disintegrates, is nothing but where the differential force of that planet on an object exceeds the resultant force on the same object. Differential force works on the sideways where as the resultant force works directly towards the planet. This is why the shoemaker-Levy revolved around the Jupiter for some amount of time even after the disintegration. If the tidal force split the comet, the pieces will dive into the planet immediately after disintegration.

Let's consider the two scenarios as shown in Fig. 3, for the resultant and differential forces.

Karunakar Marasakatla

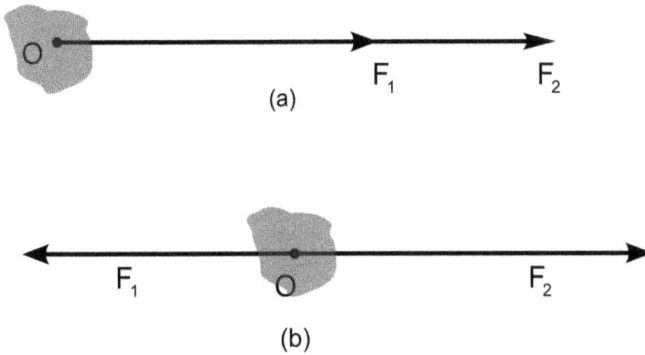

Figure 3: Different scenarios with Resultant and Differential forces

As shown in Fig 3 (a), the angle between the two forces acting at point O on an object is $0°$. The direction of the two forces is same.

Therefore, the Resultant force, R_1 = Sum of two forces

$$= F_1 + F_2 \text{ (According to the principles of statics)}$$

Differential force, D_1

$$= \text{Sum of forces} - \text{Resultant force}$$

$$= (F_1 + F_2) - (F_1 + F_2)$$

$$= 0$$

In Fig 3(b), the direction of the two forces is opposite to each other, means the angle between them is $180°$. Let's assume the two forces are equal as $F_1 = F_2$.

Then the Resultant force, $R_1 = F_1 - F_2 = 0$

Differential force, D_1 = Sum of forces – Resultant force

$$= F_1 + F_2 - 0$$

$$= F_1 + F_2$$

In these two scenarios, it is evident that the combined force of any number of forces acting on a point will be grouped into two forces, one as the resultant force and the other as the differential force. As in Fig 3

Gravity From A New Angle

(b), even though the system was in equilibrium, the differential force was acting on the object continuously putting it under stress.

Chapter 5:

What is the Mass?

Mass and its unit of measurement is a part of every day life. Gravity can be observed between two objects with a mass. There is no effect of gravity without the presence of mass. Modern physics, in fact all of the science is dependent on the definition of mass and its measurement on the surface of the earth.

Recently, it even became the subject of news when the standard mass of 1 Kg bar (International Prototype Kilogram) kept at the International Bureau of Weights and Measures located in the suburbs of Paris, France was reported as loosing its mass [6] compared to its replicas kept elsewhere in the world. So, if the mass of an object is so important in understanding the gravity and part of every day in life, what is the mass itself?

The mass was defined around three hundred years ago in the works of Isaac Newton and it is still being used in the same form as depicted below.

Mass is the measure of matter in an object. The mass of an object doesn't change if that object is heated, bent, stretched, squeezed or compressed, or transported from one place to another on earth or even to a position out in space.

According to the above definition, if we compress or expand an object, the mass of the object remains same. What is the experimental evidence for this property of the matter?

An object was never compressed or expanded as part of an experiment to check the validity of the definition of mass.

Gravity From A New Angle

A smallest shape any object could attain when compressed is a point size. First, lets explore the characteristics of this point mass object to understand more about the mass.

Point mass

Point mass object has a peculiar position in the field of physics, with having a double definition. Quantum mechanics and general relativity, two of the major theories in physics, have their own definition for this point mass object and still the concept of point mass is one of the essential part of these two branches of physics.

According to the definition of mass, which is mostly used in the gravitational theories rather than in quantum mechanics, the mass of an object doesn't change even if that object becomes a point mass.

According to the special relativity, if the matter of an object is closely packed together, then the matter will have more binding energy between the particles of matter. According to the same theory, binding energy itself is mass (the mass-energy conversion equation $E=mc^2$). It means, if an object has more binding energy then that object will measure more in mass.

So, if a group of protons and neutrons packed together closely in a nucleus, then that nucleus will weigh more than the nucleus in which the same set of particles occupied more space than earlier. A closely packed or compact nucleus will have more binding energy than the loosely packed nucleus with same number of nucleons.

At the same time, enormous amount of energy is required to compress a piece of one kilogram wood or iron bar to a point size. If all the protons and neutrons in an object occupy a point size space, then the object will have more binding energy. According to the special relativity, more binding energy in the point size wood means more mass it will measure. Therefore, the point size compressed one kilogram wood will measure more mass than the normal one kilogram wood. According to the definition of mass, both should measure the same amount of mass.

Karunakar Marasakatla

Here we see a clear conflict between the definition of mass and the mass-energy conversion principle in regard to the point mass object.

Among these two concepts, one is experimentally observed and the other is a simple imagination without any base. If we need to discard one of these two, then it would be the definition of mass.

Similar to the compact and expanded nucleus, if we compress an object, that object will measure more in gravity to the earth; therefore weighs more than the normal object. If we expand the same object, it will measure less gravity to the earth. Therefore the mass of an object, as we measure it today, does change with the size of the object.

What is the uniqueness in an object that determines the amount of mass inside it? The only distinct thing in an object is the subatomic particles it contains. Then, is the mass of that object is the total count of the subatomic particles inside it? According to the standard theory, the count of particles is not the mass of the object.

Then, is the mass of an object is the total mass of subatomic particles in it? The answer is still a no. Combined mass of all the subatomic particles in an object will be greater than the mass of the object. Then, what exactly is a mass of one kilogram of iron sphere means? Apart from saying that it is equal in comparative mass to the standard kilogram preserved at the institute, it doesn't stands for anything the object contains according to the standard definitions.

Weight, a similar term used along with the mass, was defined as the gravitational force between the mass of the object and the mass of the earth. So, by differentiating mass from weight, it was made to believe that the mass of an object is unique and doesn't depend on the gravity of earth on the object. In reality, both mass and weight are dependent on the gravity of the earth.

If we balance one kilogram of iron bar and one kilogram of loosely bound cotton in a balance scale as shown in Fig.1, then what is common in them? Everything differs in them including the size, shape, number of subatomic particles except the gravity of the earth on each of these objects. When the pull of gravity on these two objects is equal then the balance will be parallel to the surface of the earth. So the mass, we defined as one of the standard units, is in fact wholly dependent on the gravity between the earth and the object.

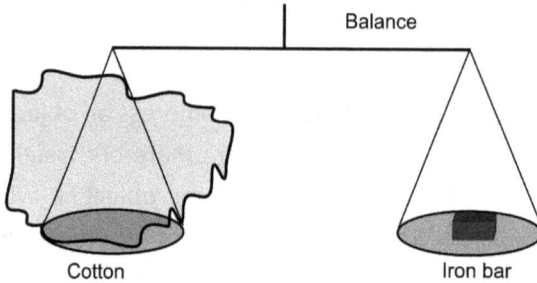

Figure 1: Cotton and iron bar in a balance scale

Therefore, there is not much difference between the mass and weight of an object according to the present definitions; both are dependent on the gravity of the earth on the object.

Therefore, if a point mass object measured more in mass means it actually measuring more gravity to the earth. Binding energy within the object has no relation to increase in the mass of the object. The compactness of the object is enabling in exerting more gravity to the earth. Therefore, point mass objects exert more gravity than their normal counterparts.

There used to be one trivia question like which weighs more, between one kilogram of cotton and one kilogram of iron? Let's explore the answer to this question. With the invalidation of the point mass definition of mass, it is now evident that the gravity differs with the size of an object. If we compress one kilogram of loosely bound cotton to the size of one kilogram iron bar, then the cotton weighs more than the iron bar. The difference might be very small, but it is distinguishable. So, answer to the question is cotton weighs more than the iron bar if we keep the size and shape of both the objects as same. Now, can we say that the mass of the cotton has increased? The only thing that changed in the cotton is its size. The size of the cotton was decreased. Then, does the mass depend on the size; can we measure the mass in cubic meters?

So the current definition of mass, as we are using it today, is flawed and not a standard one. The mass of an object keeps changing with multiple factors. It is not at all a measure for the matter inside an object. It is only a measure for the comparative gravity of the earth to the objects. But the amount of matter inside an object, the total number of basic particles, is same irrespective of shape and size of the object. Then how can we make the amount of matter inside an object as a standard or fundamental unit? Before attempting to redefine the mass as an amount of matter inside an object, let's first explore if there is anything in the nature similar to the mass as we are using it in our day to day activities.

In the nature, there are only particles and the forces between them. There is nothing like mass of an object. The strength of the force within the object is determined by how tightly the particles, atoms and molecules are packed together. In other words, the strength of the force is dependent on the density of particles within an object.

We normally assume that among the same size objects the heavier object has more mass. In fact the correct term to use for the heavier object is that it has more density of matter.

As there is no time at an instance and no distance at a point, for a single object there is no way to measure the mass within the object. Mass as we measure it today is a comparative gravity of the earth between the standard unit and the object being measured. Because gravity changes with the size and structure of the object, there is no way to determine the absolute value for the mass of an object. Therefore we can't define a standard unit for the mass to use anywhere in the universe.

We can only measure the comparative gravity of the objects to the earth on the surface of the earth at a given location. The strength of gravity of the earth to each of the objects may not remain same anywhere else on the surface of the earth or inside the earth or even away from the earth's surface.

Standard mass

Mass should represent an approximate measure of the matter inside an object. To make the mass as a measurable quantity, we need to keep both the objects being compared as a same size and shape. If the source gravity on which the measurements are made is very strong compared to the earth, then even the same size and shape objects measuring equal mass on earth might weigh differently depending upon their internal atomic structure. No two objects on earth are of same atomic structure, means the number and the position of subatomic particles are different in both the objects even if they are of same size and shape. Therefore to keep the uncertainty to a minimum, we need to define the size and shape for the standard measurement as a sphere of smallest uniform volume among all materials. We need to take that as a standard volume for the measurement of mass. Then we need to measure the strength of earth's gravity at a location on each of the material of standard volume and that will be the mass of the standard volume object for that material. This is similar to the calculation of density of an object with a difference being the smallest volume instead of one centimeter cube.

> *Standard mass of an object*
>
> *= Earth's gravity on a standard volume object at a location.*

Let's define mass index (M_i) of a material, similar to the density, as the ratio between the standard mass of the object and the standard volume.

Mass index of a material,

M_i *= (Standard mass of an object's material) / (Standard volume)*

--- Eq. 1.

Now on, the measurement of total mass inside an object will only depend upon the volume and mass index of the object.

Total mass in an object

> = (Total volume of the object) * (Mass index of the material)

> --- Eq. 2.

In this calculation of mass, irrespective of the shape of the object, the mass inside an object remains same and represents the measure of matter inside the object. The total gravity of the earth on the object may vary depending upon the shape of the object but the mass remains same irrespective of the shape of the object as long as the matter with which it was made is same. One thing to remember in this aspect is that if the size of an object changes then it should be treated as a different material and then a new standard mass has to be calculated for that material. For example the standard mass for the loose cotton and the compressed cotton has to be different as well as for water and ice.

Once the mass index for a material has been calculated, then the mass of an object made from that material should be calculated as the product of its volume and the mass index, it should never be dependent on a measurement based on the gravity again. It means, the mass of an object should never be measured in a balance scale.

Change in standard mass with the size

Let's suppose there are 1024 pieces (easy number to make half till it reaches 1 in 10 steps) of a material in object A, with each piece having a volume of 1 cm^3 and each piece measuring n units of mass according to the new definition as shown in Fig 2. For simplicity let's still call the unit as a gram and the mass of the cube as n grams.

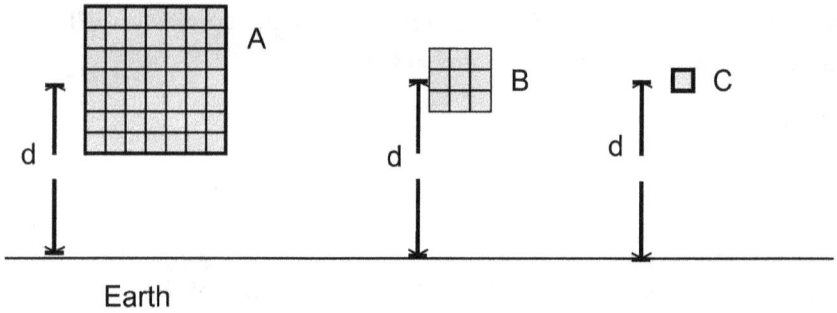

Figure 2: Change in standard mass with the size of the object

As the standard unit of measure for the mass of a material dependent on the gravity of the earth at a location, the weight of an object at the same location also depends upon the gravity. At a specific location, there is no difference between the mass and the weight of an object therefore the same unit of measure will be used to represent the total earth's gravity on an object.

Actual gravity of object A to the earth will measure less than $(1024 * n)$ grams and lets assume it as X grams. When the object A got compressed to half of its size to a size of 512 pieces of 1 cm^3 as in object B at the same location as object A then the gravity of the earth on the object B will increase and lets assume it as Y grams. When the object compressed, the standard mass of the object B will also become double from that of the object A to $(2 * n)$ grams because the density of the matter in the new object got increased. When the object A got compressed to half of its size for 10 times to a size of 1 cm^3 as in object C at the same location, its gravity to the earth will increase further more and lets assume it as Z grams. Now the Z grams is in fact equals to exactly $(1024 * n)$ grams. The standard mass of the object C will double 10 times to that of the object A.

Z grams = (1024 * n) grams > Y grams > X grams.

The gravity of these objects differs but all the objects are having same amount of mass according to the new definition of mass.

Karunakar Marasakatla

*Mass = volume * mass index*

If we take the 1 cm^3 as the standard volume, then the standard mass will be equal to the mass index of the material.

*Mass of A = 1024 * n grams*

*Mass of B = 512 * (2* n) grams*

*Mass of C = 1 * (1024 * n) grams.*

Therefore, even though all the three objects are having the same amount of matter, the measure of gravity for these objects equals to the mass in an object only when that object itself is of the size equal to the standard volume.

When the objects are bigger in size, even if it has more matter, their gravity to the earth will be less. When the same object compressed to a smaller size, all of its matter concentrates at a small volume and ultimately measures more gravity.

Change in standard mass with the distance

If we measure the gravity as the interaction between two standard volume objects then the volume can also be a visible volume of the object. As the object's standard mass increased as we compressed it, the same way, the standard mass of an object will also increase as it looks smaller and smaller from the earth. Let's take three similar objects A, E and F of each 1024 cm^3 made with same material and keep them at different distances from the earth as shown in Fig 3.

Object *A* measures the same amount of gravity, *X* grams, as in the Fig 2. The object *E*, even though it is bigger than the object *B*, it is appearing in volume same as the object *B* when viewed from a point on the earth. It means, it got visually compressed to the size of object *B*. Therefore its

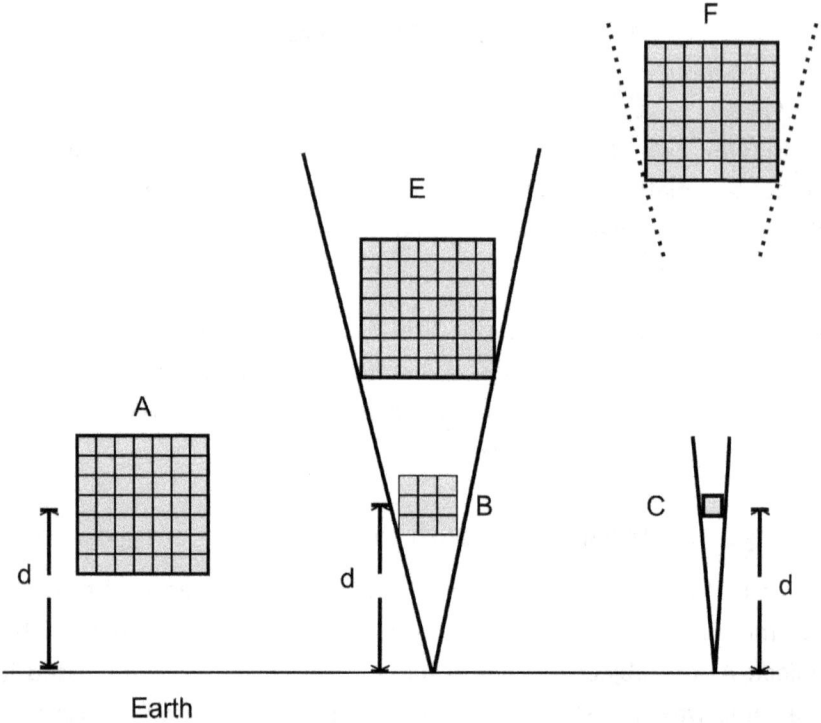

Figure 3: Change in standard mass with distance

standard mass should be measured equal to the standard mass of the object *B* as in Fig 2. In the same way, the standard mass of the object *F* should be measured as (1024 * *n*) grams because it is far away and is appearing as a cube of 1 cm^3 as the object *C* in Fig 2. Therefore the standard mass of object *F* is greater than the object *E* and the standard mass of object *E* is greater than that of the object *A*.

So, the mass as we measure it today is a comparative gravity and it changes with the shape and size of the objects. Therefore the present definition of mass should not be considered as a basic unit of measurement.

Mass of an object should be considered as a constant irrespective of a location and that should be the product of the volume of the object and mass index of the material of the object.

As the standard mass of an object increases as it compressed, the standard mass of an object will also increase if the object appears smaller from the surface of the earth.

Change in weight in different structures

Earth's gravity on an object will change when an object is compressed. Gravity will also change when different objects regrouped as a single entity.

Scenario 1: Let's suppose a pile of loose cotton was balanced with 100 kg of iron bar in a balance scale. Volume of the cotton will be much bigger than the volume of the iron bar. When the cotton is compressed to the size of the iron bar of 100 kg, the compressed cotton bar will weigh more than the iron bar.

Scenario 2: The total weight of a pile of sand grains will be less than the combined weight of each individual sand grains. If we assume the weight of a single sand grain as *x* grams then the total weight of the pile of sand of one thousand grains will be less than (1000 * *x)* grams.

As the one thousand grains compress into less volume, the weight of the pile will increase. When one thousand grains compressed to the size of a single grain then the weight will be equal to (1000 * *x)* grams.

Weight of individual proton particle is *1.672622 x 10^{-24}* grams. When two protons weighed together as a pile in the nucleus, the combined weight will be less than the total of the individual protons. This phenomenon of loss of mass in the nucleus was given a name as mass deficit.

Gravity From A New Angle

So, the definition of mass is flawed to its core. It is one of the biggest mistakes in science. Scientific community got an opportunity to correct the definition of mass when the mass difference was observed in the radioactive decay between same number of basic particles before and after the decay of elements. Instead of correcting the definition of mass, scientific community created an exception to the definition of mass and termed it as the mass deficit. If mass, the way we defined and measure, is the representation of matter then it should always be same for any given object no mater what happens with the object.

Almost every principle in physics is dependent on the definition of mass. If the definition of mass itself is flawed then all the principles derived using the mass are also flawed.

Chapter 6:

Inverse Square Law of Gravity

In Newtonian principle, force of gravity was proposed as inversely proportional to the square of the distance between the objects. If we double the distance between the objects, the strength of gravity will diminish to a one fourth of the original force between them.

A common example we find for the derivation of this principle is how

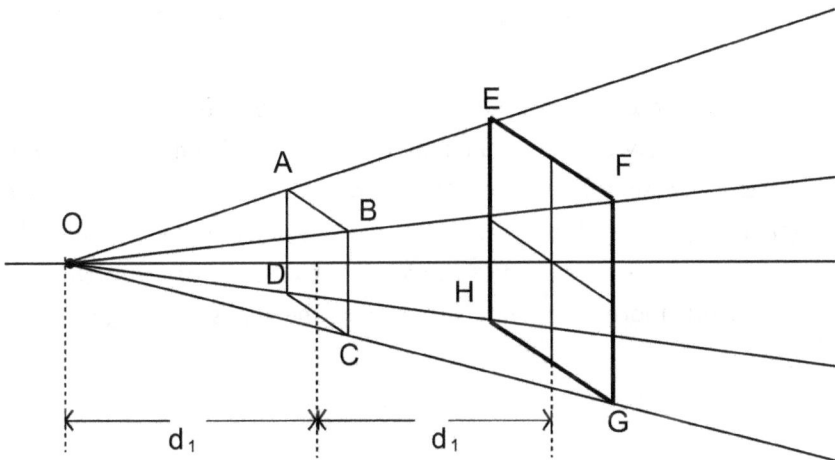

Figure 1: Inverse Square Law

the intensity of light is diminished with the increase in distance between the light emitting point and an object with an area as shown in Fig 1. O is the light emitting point and $ABCD$ is a square object.

The point O and the $ABCD$ are separated by a distance of d_1. If the square was moved away from O for a distance equal to d_1, the luminosity of O on the square at new location will decrease to one fourth of the initial.

Gravity was also assumed to work in similar to the luminosity of a light source. In the previous example, O is a light emitting object and the square is a non-luminous object. Both are dissimilar objects by nature,

Gravity From A New Angle

means one emits the light and the other doesn't. In the case of gravity, both are mutually attracting objects. If the sun exerts gravity on the earth then the earth also exerts gravity on the sun. If we compare the gravity of an object with a light emitting from an object then both the sun and the earth should be treated as light emitting sources when calculating the gravity between them.

Another assumption made when deriving the inverse square law for gravity, apart from assuming that gravity is light, is that the light source was treated as a point. In reality, wherever we study the gravity between objects, none of them were of point mass objects. Consider the example of light reaching the Mercury from the sun as shown in Fig 2.

O and S are two different points on the surface of the sun and the d is the distance between the sun and the Mercury from point O. The distance between the sun and the Mercury from point S is x. If we double the distance of the Mercury from the sun from point O, the light reaching the Mercury from the sun wouldn't be $1/4^{th}$ of the original, it will be slightly more than a quarter because the sun is emitting the light

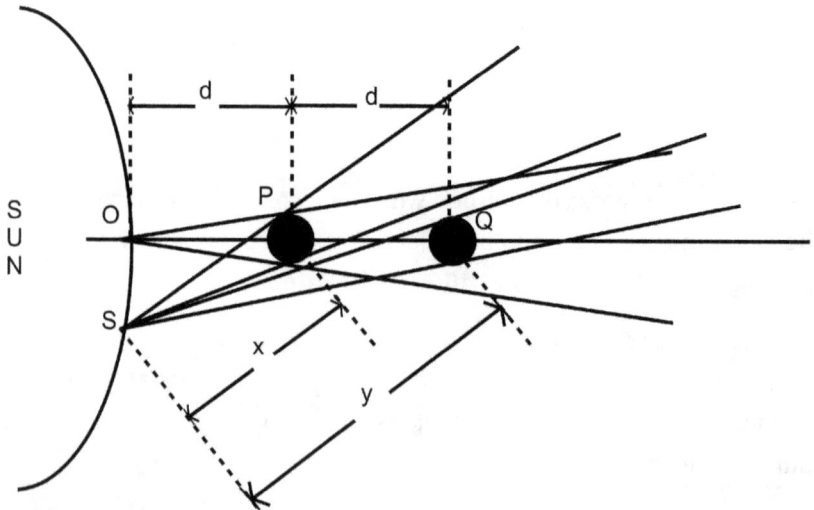

Figure 2: Relevance of Inverse square law for the distance between Sun and Mercury.

Karunakar Marasakatla

in a wide angle instead of as a point. Even the new position of the Mercury wouldn't be exactly double from all other points on the sun. When the new position of the Mercury viewed from point *S*, the new distance *y* is not double the distance of the earlier distance of *x*. Therefore more light reaches the Mercury than $1/4^{th}$ of the original.

As seen earlier in the chapter on mass, an object and its point mass counterpart are not same. When a bigger object turns to a point mass object, the new point mass object weighs more. According to the present definition of mass, the new point mass object gains more mass. Therefore, any principle derived assuming these two objects are same is invalid.

Gravity is not a thing that emanates all around an object like the light emanating from a light source. Light source emanates the light without any external influence. Gravity is a mutual entity. It will be felt on an object only when there is another object in its vicinity. If there is only one object, there wouldn't be any effect of gravity around it.

We see the waves in the ocean generated by the gravity between the earth and the moon travelling all around the earth due to the rotation of the earth and the revolution of the moon around the earth. If the earth and the moon remains stationary at a distance then the waves in the ocean always point towards the direction of the moon.

Another aspect which inverse square law ignored is the differential force of an object like planet Jupiter. Bigger objects will have both resultant and differential forces. A point size object will have only the resultant force. The effect of differential force is missing in the point mass object.

Gravity doesn't act like the light and the objects gravitationally interacting are not point mass objects. Therefore the inverse square law principle, which was derived based on the characteristics of the light emanating from a point source, wouldn't be valid for the gravity.

The Inverse square law principle is widely applied in physics in the derivation of the strength of light, sound, electricity and magnetism at a new location away from the source. It is valid for the propagation of light and sound but not for the gravity. Inverse square law is valid for the scenarios where the strength of a force acts all around the object irrespective of existence of another object around it.

Chapter 7:

Prevailing Theories of Gravity

Newtonian theory of Gravity

Newton formulated most of the principles in mechanics such as the definition of mass, force, resultant force and the inverse square law of gravity. Newtonian theory of gravity is wholly dependent on these flawed concepts. Inverse square law is part of the principle of gravity. When the inverse square law itself is invalidated, the principle of gravity, $F = Gm_1m_2/d^2$, automatically gets invalidated.

Newtonian gravitational principle was able to describe the nature of the gravity in the solar system. It successfully incorporated all the observed data at that time and also appeared to have predicted the existence of new planets. Newtonian principle is a convenient fit of the observed data. This doesn't describe the true intrinsic property of the matter, the gravity.

Newton derived a principle for the observed gravity between two spheres, earth and the moon, and applied that to point mass objects, assuming that both will behave in the similar manner gravitationally.

The conflict between the quantum mechanics and gravitational theories about the point mass object also proves beyond doubt that the Newtonian principle of gravity is not the true nature of gravity.

General Relativity

General relativity is the widely acclaimed theory of gravity among all other theories. It has been described as the standard theory of gravity. But still, it is not completely accepted in the scientific community [7]. There were many attempts to explain the gravity in alternate theories but none of them were able to dethrone the general relativity.

The opposition to General Relativity is basically stemmed from the notion that it is still in odds with the other standard theory of particle dynamics, the quantum mechanics. And also, there is no practical application evolved from this theory in it's almost a century of existence. This theory has been used to explain the cosmological events but failed to absorb the facts observed at the quantum level. Mass of an object is a major component in any theory of gravity. We have conveniently derived the mass of all the objects in space using the gravitational theories and applied the same mass back to the same theories. There is no other way to confirm the mass of the planet derived using these principles. Lack of applicability and no logical way to disprove the theory made many people to wonder the completeness of the theory.

Secondly, it is very difficult to visualize the dent or curvature in space around massive objects as proposed by this theory. A common example given for the curvature in space is the dent created by a massive object when it was placed in the middle of a stretched out sheet. In similar to the energy, we can't create or destroy the space. If we gain space on one side of the object, the space will be reduced by same amount on the other side. If we have a dent in the sheet on one side of the object then the other side will have a bump, ultimately loosing the space on that side. If an object circles around another massive object when it approached the massive object on the side where there was a curvature, the same object will get repulsed if it approached the massive object from the opposite side of the curvature. This is not happening in the nature; therefore the space is not curved around any object.

Gravity is not a curvature in space. It is a force with which all the particles attract each other. General relativity basically failed at incorporating the grand picture of the universe. It completely ignored the forces acting in the quantum world. From the time it was proposed, there were many new observations in the cosmos which were not explained by the general relativity, for example, the Pioneer anomaly and the orbital speeds of the stars on the outskirts of the galaxy.

The conflict between the special and general relativities regarding the energy content of a compressed point size object is a clear indication to the flaws in the principles of general relativity. If we compress an object like a iron sphere of one meter radius to a point size, it will have more binding energy than earlier. More binding energy in a object is more mass it measures. The same object will measure different amount of mass depending upon the compression level of the object. The dependence of general relativity on the flawed concepts of mass and the inverse square law of gravity automatically invalidates the theorem.

Albert Einstein formulated the special relativity earlier than the general relativity. Surprisingly, special relativity contradicts the definition of mass and the general relativity is wholly dependent on this principle. No wonder why special relativity has many applications and the general relativity has none.

Modified Newtonian Dynamics

MOND or as it stands for Modified Newtonian Dynamics is a proposal to change the Newtonian principle for the acceleration to apply to the objects at greater distances[8,9].

Newtonian principle for the force is as follows when equated with the acceleration -

$$F = ma.$$

The modification proposed to the above principle to apply at greater distances in the MOND is

$$F = ma^2.$$

In essence, the modification explains that more force is not required to keep the objects in an orbit where the acceleration is small.

Because the notion of mass itself is flawed in present theories, any theory that depends on the present definition of the mass will also gets invalidated.

The above Newtonian principle for the force itself is not valid; therefore any other new principle derived from that equation will also be invalid.

It is well known that the current theories are unable to explain the anomalies at greater distances like the galactic rotation curve and the pioneer anomaly. Asking for to change the principles at greater distances without any basic explanation is irrational.

So far we have discussed about the flaws in conventional theories and resolutions for these misunderstandings in the science. Conventional theories about the force, work and the resultant force, the concept of fixed mass for a given object and the inverse square law of gravity are basically impeded the understanding of the nature, in particular the mother of all forces, gravity. In the rest of the book we define the gravity and its applications.

Part II: New Theory of Gravity

It requires energy to displace a resting object in any direction. A force or gravity can displace an object from its resting place. Therefore, gravity is nothing but a force and its source of energy is continuously being consumed when it does a work.

Gravity can be defined as a force between two point mass objects. It is an intrinsic property of the matter. The force we see in the nature between two non-point mass objects is the resultant force of the intrinsic gravity of all the point mass particles of one object to all the point mass particles of another object.

Gravity appears to be very strong force and in fact the strong nuclear force is nothing but the gravity in its intrinsic form between the subatomic particles.

This section defines the new principle for the gravity and explores the different scenarios for the strength of gravity between different objects with different size, shape and orientation towards each other.

Chapter 8:

What is Gravity?

Gravity is simply the energy we see all around in the nature. Energy converts from one form to another in a cyclic manner. Vibration, sound, heat, light, x-rays and all the radiation generates at the expense of the gravity. It is also possible to convert all these forms of energy back into the gravity.

Consider the following scenario where a chemical powered battery generates electricity and the electricity generates heat in a coil. Thermal energy generated by the coil was used to melt an iron powder into a solid bar. Iron powder weighs less and the same material when converted into a solid bar, weighs more. It means, the powder when converted into bar, gained more gravity because the size of the object got compressed. In essence the gravitational self energy, the total gravity between all the particles of the object got increased. Here the chemical energy was converted into electricity, electricity into thermal energy, and thermal energy into gravitational energy.

If the solid iron bar converts back into the form of powdered iron on its own, it will release the gravitational self energy. The released energy can be converted into thermal energy. Thermal energy can be converted back into the electrical energy and ultimately the electrical energy can be converted into the original chemical energy by forming the original chemicals involved in the chemical reactions.

Energy is in many different forms. We can convert the energy from one form to another but can't create or destroy the energy. Even though the energy is in many different forms, the underlying nature of all these forms of energy is the gravity. Gravity was assumed as something different from what we see everyday. It turned out to be everything what we see in each and every movement.

Gravity can be defined as a force between two point mass objects. It is an intrinsic property of the matter. The force we see in the nature between two non-point mass objects is the resultant force of the intrinsic gravity of all the point mass particles of one object to all the point mass particles of another object.

Gravity doesn't work as light emanating from a light source, like the light spreading from the sun all around it. Also, for an object to exert gravity there has to be another object in its vicinity. It is a mutual interaction. Gravity also works only on the line of action, not all around the object. The strength of gravity is not inversely proportionate to the square of the distance between the objects.

Chapter 9:

Principle of Gravity

The force of gravity can be observed between all the objects of matter regardless of their mass or size. The strength of gravity depends upon the mass inside the objects and their size as well as the distance between them. If the object is a non sphere object, then the gravity is also depends on the position or orientation of the object in reference to the other object.

To arrive at a simplistic form of gravity, we need to make the gravity dependent on less number of factors. Two point mass objects have no shape, size or orientation towards each other. Because gravity works at all objects of the matter, it is better to use point mass objects to define the gravity between two objects.

Let's assume two point mass objects with masses as m_1 and m_2. Point mass objects wouldn't have any shape or size; therefore there wouldn't be any issue with the orientation of the objects to each other as well.

Intrinsic Gravity

Intrinsic gravity between two point mass objects is proportional to the product of the mass of the objects and inversely proportional to the distance between the objects. This can be formulated as follows -

Intrinsic Gravity, $F \alpha\, m_1 m_2 / d_{12}$

m_1, m_2 - mass of the point mass objects

d_{12} - distance between the objects.

When it is equated on both sides with a new constant K, then the formula will be

$F = K m_1 m_2 / d_{12}$ --- Eq. 1.

Constant K is the new gravitational constant, different from the Newtonian or universal gravitational constant, G.

Gravity is a mutual attraction between two objects, acting on the line connecting these objects. If we separate the two point mass objects, keeping their individual mass as constant, the force between them will decrease linearly.

Newtonian principle concentrated on the resultant force between the objects, whereas the new principle defined the intrinsic gravity between them, from which the resultant force between them can be deduced.

Intrinsic gravity F, is the energy generated between the objects in a second.

Gravity between non-point mass objects

In the nature, nothing is a point mass object. In that case, how can we measure the gravity between bigger objects than the point mass objects.

Gravity between two objects is the combined gravity between each of the point masses in one object to all the point masses in the other object. It is a many to many scenario.

As discussed in the chapter on mass, there is no way to measure the mass inside an object precisely. We can only measure the mass in an object comparative to the standard volume defined in the chapter on mass.

Then the mass of the object will be

*Mass of the object = Volume of the object * Mass index of the material*

$$m = v * M_i$$

Where M_i = standard mass / standard volume.

As the standard volume tends towards the volume of the point mass object, we get more accurate measure for the mass of the object.

Karunakar Marasakatla

For non point mass objects, to calculate the gravity, the object has to be divided into standard volume parts. The standard volume parts in one object should be in such a way that when viewed from another object, the standard part in the first object should look like a point mass.

Gravity between two standard volume parts in different objects is

$$F \alpha (sv_1 * sm_1 * sv_2 * sm_2) / (sv_1 * sv_2 * d_{12})$$

Where, sv_1 = standard volume of the object 1

sm_1 = standard mass of the object 1

sv_2 = standard volume of the object 2

sm_2 = standard mass of the object

d_{12} = distance between standard volume part in object 1 to the standard volume part in object 2.

If we eliminate the common factors,

$$F \alpha (sm_1 * sm_2) / d_{12}$$
$$F = (C * sm_1 * sm_2) / d_{12} \qquad \text{--- Eq. 2.}$$

Gravity between the objects is the combined gravity or the resultant force of all the gravity between all standard volume parts of one object to the other.

As the standard volume tends to zero, the value of C approaches the value of K.

Gravity between two distant stars

Interesting thing about point mass object is that it has to be visible to other object as a point, not necessarily in a physical sense as a point particle.

When viewed from earth, a star is a point mass object even though it is thousand times bigger than the earth. Even the galactic center at the center of our Milky Way galaxy should be considered as a point mass when viewed from the earth.

As discussed in the chapter on mass, when a number of particles collapse into a single point, all the combined mass of the individual particles will be the mass of the new point mass particle or object.

Therefore the gravity between two distant stars will take the intrinsic form of gravity.

Resultant Gravity

As explained earlier, the gravity we observe is the resultant force of the intrinsic gravity between all the particles of one object to all the particles of the other object.

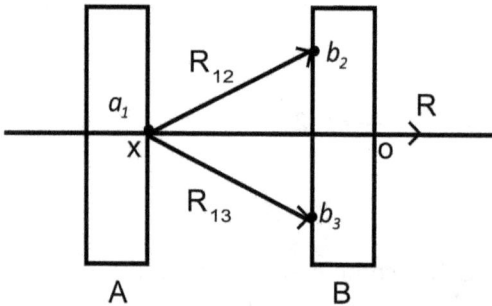

Figure 1: Resultant gravity

And also as discussed in the chapter on resultant force, when multiple forces interact at a point, there will be a resultant force as well as differential force. When we apply this principle to the forces of gravity, we get

Combined or total Intrinsic Gravity =

Displacement Gravity (or Resultant Gravity)

+ Differential Gravity.

$$I_g = R_g + D_g \qquad \text{--- Eq. 3.}$$

To explain it further, lets explore the gravity between a single point particle on a bar object A to the two point particles on another bar object B as shown in Fig. 1.

The particles a_1, b_2 and b_3 are the point mass particles on the objects A and B respectively. m_1 is the mass of the point particle a_1 and m_2 is the mass of the point particles b_2 and b_3. The particles b_2 and b_3 are at the same distance, d from the particle a_1. Lets assume that the θ is the angle b_2 and b_3 makes at the particle a_1.

According to the new theory of the gravity, the resultant gravity of intrinsic gravity between a_1 and b_2 (R_{12}) and the intrinsic gravity between a_1 and b_3 (R_{13}) works along the line going through the points x and o.

$$R_{12} = Km_1m_2/d, \ R_{13} = Km_1m_2/d$$

The resultant gravity, R of the intrinsic gravity forces, R_{12} and R_{13}, acting on the particle a_1 with an angle of θ is -

$Resultant \ Gravity, R = sqrt \ (R_{12}{}^2 + R_{13}{}^2)$

$$= sqrt \ (R_{x\text{-}component}{}^2 + R_{y\text{-}component}{}^2)$$

$$R_{x\text{-}component} = R_{12\,x\text{-}component} + R_{13\,x\text{-}component}$$

$$R_{y\text{-}component} = R_{12\,y\text{-}component} + R_{13\,y\text{-}component}$$

If we align the force R_{13} with the x-axis, the x-component of the force will be equal to the force itself and the y component of the force will be zero according to the principles of statics.

$$R_{13\,x\text{-}component} = Km_1m_2/d$$

$$R_{13\,y\text{-}component} = 0$$

Then the x and y components of the force R_{12} will be

$$R_{12\,x\text{-}component} = (Km_1m_2/d) * Cos \ \theta$$

$$R_{12\,y\text{-}component} = (Km_1m_2/d) * Sin \ \theta$$

Then, $R = sqrt \ (\ (R_{13\,x\text{-}component} + R_{12\,x\text{-}component})^2$

$$+ (R_{13 \ y\text{-}component} + R_{12 \ y\text{-}component})^2)$$

$$= sqrt (((Km_1m_2/d) + ((Km_1m_2/d) * Cos \ \theta))^2$$

$$+ (0 + ((Km_1m_2/d) * Sin \ \theta))^2)$$

$$= (Km_1 \ m_2/ d) * sqrt ((1 + Cos \ \theta)^2 + Sin^2 \ \theta) \qquad \text{--- Eq. 4.}$$

As the angle θ decreases, the combined gravity will increase. When the angle is 0°, i.e. when b_2 and b_3 were on the line x and o, means when they merge into a single point then the gravity will be equal to the sum of both the forces of gravity, a maximum value for the force between them. As the θ increases, the combined gravity will decrease. When it is maximum, i.e., 180°, the combined resultant gravity will be zero. In this scenario, the differential gravity will be equal to the total intrinsic gravity. We can only achieve the wide angle 180° when an object encircles the other object.

Therefore the angle, the particles of one object makes on the particles of other object, is important in determining the gravity between two objects. In essence, the visible size and shape of the objects is very important in determining the gravity.

Gravitational Self Energy (GSE)

Gravitational self energy is the sum of all intrinsic gravity between each particle of the object to all the particles of the same object. In other words, it is the strength of all the particles of the object with which it was bound together. If all the particles in the object are tightly bound together, where the distance between the particles is less, it will have more gravitational self energy and measures more gravity on a planet. If another object has same number of particles as the above but they are placed away from each other then the gravitational self energy of that object will be less and it will measure less gravity on the same planet.

Basically, the gravitational self energy of an object means the energy required to split each of the particles away from each other.

Gravitational self energy of the water is more than that of the ice formed from the same amount of water because the water occupies less space than the ice. Even though water has more gravitational self energy, it doesn't appear to be stronger than the ice because the water molecules have more kinetic energy than the ice which has zero kinetic energy in the molecules.

Because gravity is able to split apart the object in the form of differential gravity, it is essential for an object to have lot of gravitational self energy to be intact under a stronger gravitational force. If a pile of ruble, like a loosely coupled asteroid approaches a bigger planet, the differential gravity of the planet will rip off all the rubble from the asteroid in the initial encounter itself. Therefore, if an object has more gravitational self energy then it exerts more gravity on other objects; at the same time it sustains the differential gravity of the other objects. Loosely bound objects will remain intact if they revolve around the sun at the edge of the solar system. At that distance, the differential gravity as well as the resultant gravity of the sun will be less.

Chapter 10:

Shell Theorem

Definition of mass, inverse square law of gravity and the shell theorem are core part of the gravitational theories. Shell theorem was derived from the inverse square law of gravity using the current definition of mass. As we have seen earlier, the inverse square law and the current definition of mass are flawed. When a dependent parameter is flawed,

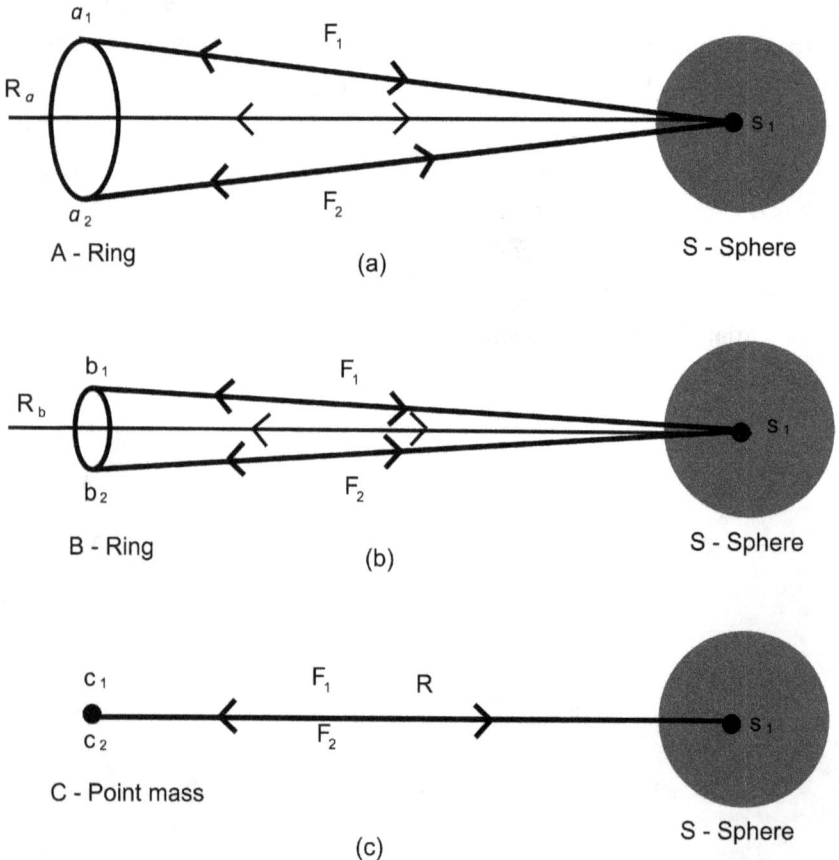

Figure 1: Gravity between a ring and a point in the sphere

the whole theorem needs to be re-evaluated.

There are two parts in the shell theorem. The first part of the theorem states that the mutual gravity between two spherical objects separated with a distance doesn't depend upon the size of the objects.

The second part of the theorem states that the gravity of a shell on an object within the shell is zero irrespective of its position inside the shell.

When we apply new definition of mass and the new principle of gravity, which is inversely proportional to the distance between objects to these scenarios, we get different conclusions.

For the first scenario, the gravity between two spherical objects does depend upon the size of the objects. Gravity between these two objects will be at maximum when these two objects become point mass objects at the center of the original objects.

In the second scenario, the gravity between a shell and an object inside the shell will be a non-zero value. The gravity will be equal to zero only when the object is at the center of the shell. Any object close to the wall of the shell will be pulled towards the center of the shell.

Following are the conclusions of the shell theorem using the new principle of gravity and the new definition of mass.

- Intrinsic or Total Gravity of a sphere object on an external point mass object remains same irrespective of the size of the sphere.
- When the size of the sphere is decreased, resultant gravity will increase and the differential gravity will decrease.
- For a point mass object, the resultant gravity and the intrinsic gravity becomes equal. Differential gravity for a point size object will be zero.
- Resultant gravity inside a shell is a non-zero value except at its center. The resultant gravity will pull any object inside the shell towards its center.

- Resultant gravity at the center of a shell will be zero. Therefore the differential gravity will be equal to the intrinsic gravity between the shell and the object inside the shell.

All the experience and the observation tell us that the combined force will be more when the forces acting at a point were close to each other.

Consider the gravity between the two particles a_1 and a_2 of a ring object A to a particle s_1 of a sphere S as shown in Fig 1(a). F_1 is the gravitational force between points a_1 and s_1. F_2 is the gravitational force between points a_2 and s_1.

If the ring object A was compressed to the size of a ring object B as shown in Fig 1(b), then the angle the particles b_1 and b_2 of ring B make at s_1 will decrease. Assume that the mass of all particles a_1, a_2, b_1 and b_2 are same. Then the resultant force of the forces F_1 and F_2 will be more compared to when the radius of the ring was greater as in Fig 1(a). R_b is the combined resultant force of all the point pairs on the ring object B to the point s_1 on the object S. Because the angle is less, the resultant force R_b will be more than the resultant force R_a.

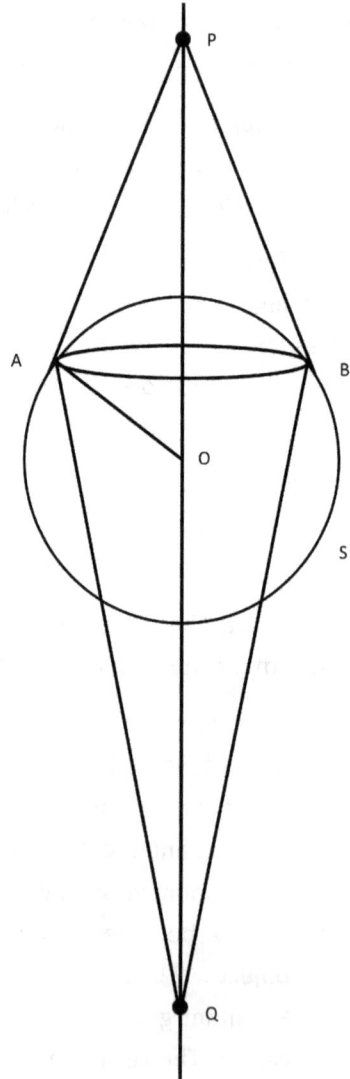

Figure 2: Gravity between a shell and a point (shell theorem)

If the ring object B is compressed to a size of point mass object C, means the angle the points c_1 and c_2 make at the point s_1 is zero as shown in Fig 1(c), then the resultant force of F_1 and F_2 will be equal to the sum of forces. In this case, not only the point pairs of c_1 and c_2 but also all other pairs on point mass object C make a zero degree angle at the point s_1 of object S. This dramatically increases the combined resultant force, R between the point mass object C and the point s_1 on the sphere object S. Even though C itself is a point mass object, the point pairs were assumed to be on the object to demonstrate the similarities in all these scenarios.

Therefore, the gravity between objects will increase as the objects compress into a smaller volume. If the sphere S is also converted to a size of point mass then the total force between the two objects will be of a maximum value for that distance. In similar way, the gravity between objects will decrease if the object expands in size.

So, if we break a compact iron sphere into small pieces, the rubble of pieces would weigh less than the initial sphere if the rubble occupies more space than the sphere. In the same way, if water expands to become the ice, ice weigh less than the weight of the initial water. The difference in weight might be very small, but it is obvious.

Apart from these drawbacks, there are some other obvious flaws in the shell theorem.

In the derivation of the shell theorem, the sphere was assumed as a concentric shells and each shell was assumed as a set of bands or rings. In the Fig 2, S is a shell and the AB is a band or ring on the shell. When the shell is sliced along the line PQ, we get a set of bands or rings around the line PQ. AB is one of such band on the shell. P is the external point mass object. The gravity between the ring and the point mass was derived and then aggregated for all the rings in the shell to get the combined gravity of the shell to the external point mass object.

The gravity between the ring and the external point mass object was calculated as the vector sum of all the forces of the ring to the external point mass. According to the present theories in physics, if the earth becomes a point mass at its center, there wouldn't be any change in the gravity of the earth on the objects around it. Then let's assume that the earth is also a point mass, *Q* as shown in the Fig. 2.

In calculating the gravity between the ring and the point earth, the total gravity was not considered as the vector sum of all the forces between the point earth and the ring. It is because we defined the mass as an entity which doesn't depend on the size of the object. The present theories states that the mass of an object is evenly distributed in whole body of the object. Later on, the mass of any part of the sphere either it as a shell or a ring of the shell is defined simply as the product of volume and the density. In other words the shell theorem itself is in the form of definition of mass. Even before proving the shell theorem, we used it in its own derivation in the form of mass and proved an imaginary theory. Mass, as we have seen earlier, is a measure for the strength of gravity between the object and the earth. Therefore the mass of the ring can't be defined as the product of density and the volume, it is a vector sum of the forces to the external entity, either it could be a point mass or point earth.

All the theories in physics are the approximation of forces at different levels with objects of different sizes. There is no relation between one theory to another. When we discard the imaginary definition of mass and the baseless inverse square law and the derived shell theorem, the entire science appears in a unified way.

Gravity between the massive earth and an iron bar on the surface of the earth appears to be very small compared to the large mass of the earth because most of the mass of the earth is widely distributed in an angle of 180° on the surface of the earth. The resultant force of all the particles of the earth on the particles of an object on the surface of the earth would be very small. If all the mass in the earth concentrates at the center of the earth, the weight of the same iron bar, at the same

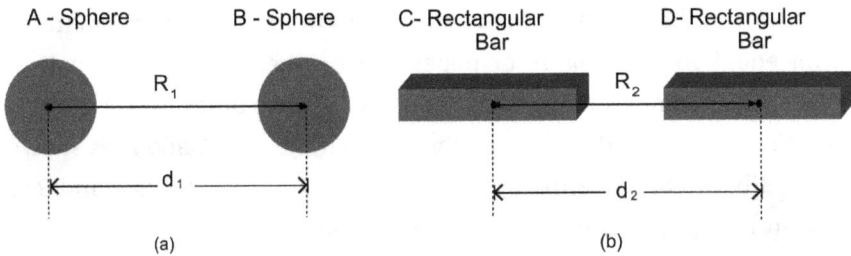

Figure 3: Gravity between spheres and bars

distance from the center of the earth, will increase many fold.

All these scenarios prove that the shell theorem is flawed. Therefore the gravity does depend on the size of the objects.

Equivalence Principle

Equivalence principle states that the acceleration of an object due to gravity wouldn't depend upon the mass of the object.

As the current definition of mass is flawed, the relevance of matter inside an object to the acceleration of that object needs to be explored further. If we keep the amount of matter inside an object as a constant, then the compressed object of that matter will have more gravity than the expanded object. Therefore the compressed object will have more acceleration towards the earth than the expanded object of the same matter.

Therefore the acceleration of objects towards the earth wouldn't be same for all objects. Equivalence principle will fail if tested with different objects of same matter with different sizes.

Does gravity depend on the shape of the object as well? Yes. The gravity does depend on the shape of the objects as well. The angle made by two particles of one object to a particle on another object does depend on the shape of the objects.

In the Fig 3, the gravity between two different objects of same material with equal volume will differ depending on their shape. Each pair of objects is of same distance from the center of each other. All the objects are also of uniform density. Gravity between the bars C and D as shown in Fig 3(b) will be more than the gravity between the spheres A and B as shown in Fig 3(a). The distance d_1 and d_2 are equal but the R_2 will be greater than the R_1.

Gravity also does depend on the orientation of the objects to each other. In Fig. 4, the gravity will increase from left to right even though

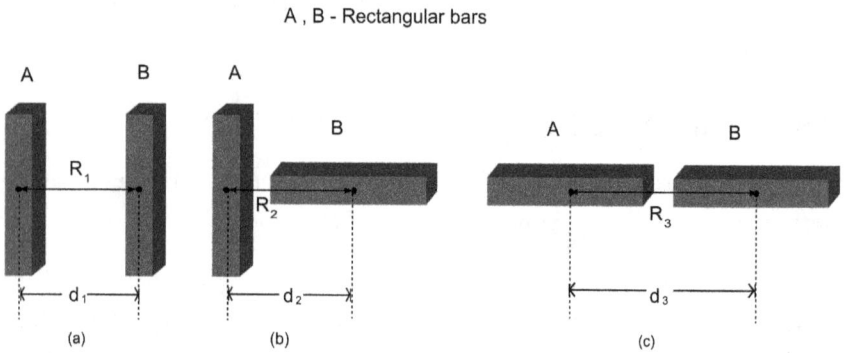

Figure 4: Gravity between identical bars with different orientation

the objects and their distances from each other are same in each scenario.

The distances between the bars d_1, d_2 and d_3 are equal. When the bars are in vertical position as in Fig 4(a), the particles on one bar make a wide angle at any of the particle on the other bar, therefore the gravity R_1 will be less between the bars in this scenario.

In Fig 4 (c), the particles on one bar makes a small angle on the particle of the other bar, as a result the gravity R_3 will be more in this scenario. The gravity R_2 will be less than the R_3 but greater than the R_1.

Chapter 11:

How the Gravity works?

Gravitationally bound system will continuously expend the energy. Earth is using its gravitational self energy to continuously pull the moon towards its center. Because the moon has kinetic energy, it moves forward in a straight line. The resulting force from the earth's gravitational force and kinetic energy of the moon keeps the moon in the orbit around the earth.

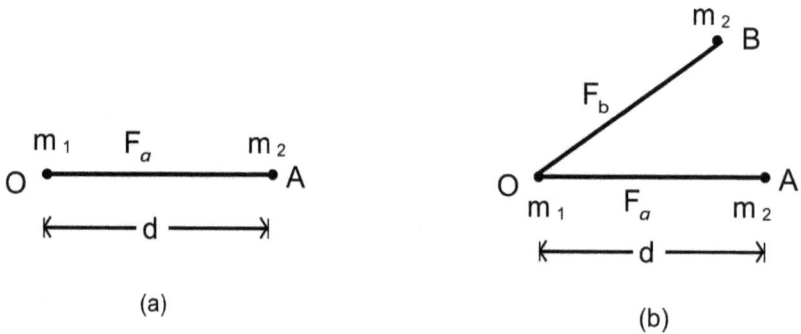

Figure 1: Gravity between two and three point mass objects

Gravity is ubiquitous. The consumed energy is spread all over the universe. It is nothing but the radiation that is filling the universe including the visible light we see all around. To know the workings of the gravity lets consider the following scenario as shown in Fig 1 (a) and (b).

As shown in Fig 1 (a), O and A are two point mass objects with a mass of m_1 and m_2, separated with a distance of d. The gravitational force F_a works on the line connecting the two objects. Now let's add another point mass object B to the system with a mass of m_2, equal to the mass of the object A as shown in Fig 1 (b).

Object O also attracts the object B with same gravitational force as the object A. In this case, the F_a will be equal to F_b. For a movement, let's ignore the gravity between the objects A and B.

Now lets add few more point mass objects with the same mass as m_2 around the O with same distance as d as shown in Fig 2 (a).

All the forces of objects A, B, C, D, E, F, G and H are equal on the object O. Now, assume that all these point mass objects are part of a ring M with a width of a single point mass particle around the point mass object O as shown in Fig 2 (b). Assume that all these particles on the ring were tightly bound together as a single object.

Mass of the ring M is equal to the total mass of all the point mass

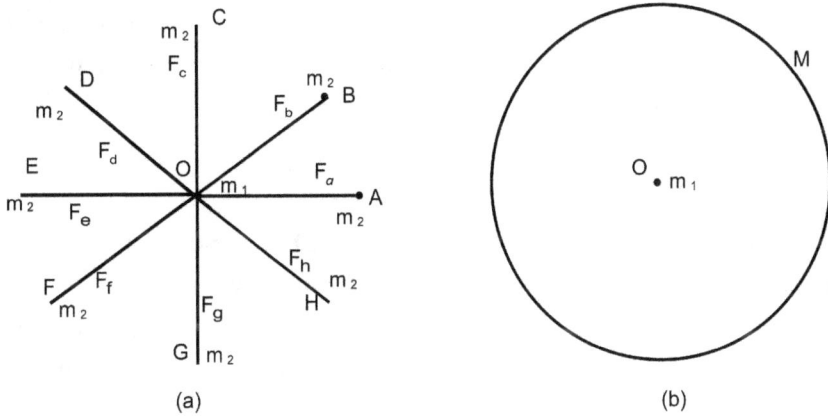

(a)   (b)

Figure 2: Gravity between single point mass and multiple point mass objects

particles on it, each having a mass of m_2. Total intrinsic gravity exerted by the ring on the object O is the total gravity of all the particles on the ring.

If an object is pulled from all angles, it will break apart after certain time. How long it sustains depends upon its internal strength. Now the question is how much gravity the object O can sustain? It is equal to the total internal energy of the object O. When the mass of the ring is increased by increasing the mass of individual point mass objects, at one point of time, all the internal energy of the object O will be

Gravity From A New Angle

released. If the object at the center is a big object then it will break into small pieces. If it is an atom, the nucleus of the atom will disintegrate.

Gravity is a continuously acting force in the nature. Current theories states that the gravitationally bound system will never change. It was assumed that a gravitationally bound system doesn't expend any energy at all. Gravitationally bound system does expend energy. The energy comes from the internal self energy of the objects.

If an object like the earth revolves around a central object, like the sun, the central object looses equal amount of energy required to keep the earth in orbit. If many objects revolve around the central object, like the solar system, the central object will loose tremendous amount of energy. In fact, the loss of internal energy itself makes the sun, the central object to glow. A dim star is without a planetary system or any other material around it. A bright star is with many planetary bodies revolving around it.

How the Sun shines?

It is evident that the core and the rest of the body of the sun rotate at different speeds. From this difference in rotation, we can deduce that the core and the rest of the body of the sun are decoupled at the boundary. It is also widely known that the core of the sun was made of a dense material than the rest of the body.

Gravity will be present when two objects are separated by a distance; therefore the decoupled outer part of the sun and the inner core are gravitationally bound together. The outer layer revolves around the core just like another planet. Because the outer layer revolves around the core, they both have different rotational speed.

Outer part of the sun and the planetary bodies exerts the force on the inner core; as a result the inner core releases the energy in the sun. The inner core releases the energy either as it expands or as the material separates from the core into the outer plasma layer. The gravity of the inner core will decrease as it expands or as it looses the material. The outer layer of the star escapes the star in an explosion, the so called supernova, when the inner core is no longer exerts enough gravity to hold the outer layers.

Plasma might as well release the energy when it converts into the element helium but the bulk of the energy will be released from the gravitational self energy of the core of the star.

The chances of dense core of the sun being a neutron core are more [10] as shown in Fig 3. As the neutrons are released from the core, they interact with the plasma to form the element helium.

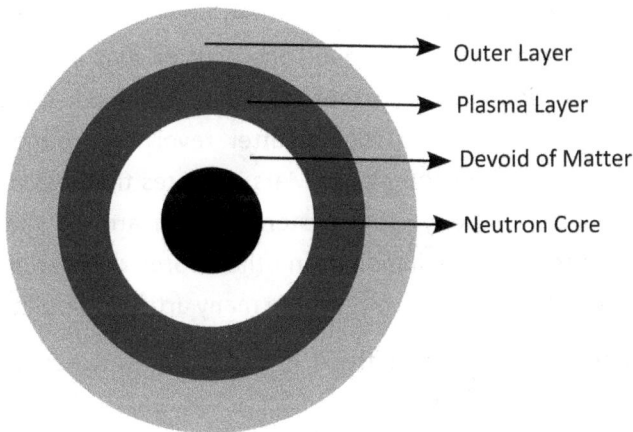

Outer Layer

Plasma Layer

Devoid of Matter

Neutron Core

Figure 3: Structure of the Sun

Each planet with a satellite should also radiate energy. We can call the Saturn as a failed star instead of the Jupiter because it has more satellites or more material revolving around it. A few more satellites would have made it brighter. The mysterious radiation from the planets with satellites might be the loss of energy in holding the satellites around them.

Young stars are dim because dust particles distribute the energy all along the star. Once the dust particles coalesce as planets; the force exerted on the star at one point of time increases. This makes the stars to shed the cocoon and radiate the internal energy.

A star within a dust nebula will be boiling within a cocoon. Star with planets will have convection cells which radiate energy. This makes the star brighter. The distance between the core and the plasma layer might have also influenced the ignition of a star. As the star contracted from the dust nebula, the distance between the core and the plasma might have also reduced. When the gravity between them increased, the star might have ignited by releasing more energy.

A star under extreme gravitational pull radiates high energy radiation like the x-rays. The energy radiating from an object is an indication to how much matter is revolving around the core of that object. Strong high energy radiation similar to the one emanating from the galactic center indicates that there is more matter revolving around it, less radiation like the earth or the planet Mars indicates that less matter is revolving around that body. The mysterious hum around the planet earth might be the same radiation or in other words the loss of energy due to the revolution of the moon and the many artificial satellites.

Current theories proclaim that there is no gravity between a ring of point mass objects and a point mass object at the center of the ring. The same applies for a point mass at the center of a shell.

Ring of particles has mass as well as the particle at the center as in the Fig 2(b). If there is mass, there will be gravity. When we apply the scenario of ring and particle to the new theory and the new principle of gravity, we get a gravity of more than zero depending upon the mass of the objects and distance between them (radius of the ring or shell).

Total Intrinsic Gravity = Displacement Gravity (Resultant Gravity)

+ Differential Gravity

Because the point mass particle is at the center of the ring of particles, the angle any of the two opposite particles on the ring make at the central object will be 180^0. Therefore the displacement or resultant gravity will be equal to zero and the differential gravity will be equal to the sum of intrinsic gravity between the objects. If the ring is strong enough, the differential gravity between the ring and the particle will pull apart the particle at the center.

When two objects are bound together by the gravity, where the energy comes from to keep the system in equilibrium?

If two objects are approaching each other because of the gravitational attraction, the energy consumed in the system comes from the gravitational self energy of both the objects. If the two objects bound together by one object revolving around another object, the energy consumed comes from the gravitational self energy of the object at the center. Gravitational self energy of the object at the center will be used to keep the revolving object with kinetic energy in the orbit. The energy released will radiate from the central object.

As the central object continues to loose its gravitational self energy, its gravity on the revolving object will decrease. Eventually the system will expand.

Chapter 12:

Gravity between a Ring and a Point Mass

Gravity depends on lot many factors than just the mass and distance of the objects. Each scenario with different shapes of objects will have a unique equation for the strength of the gravity between objects.

In this chapter, we will derive an equation for the gravity between ring of point mass particles and a point mass particle on the line going

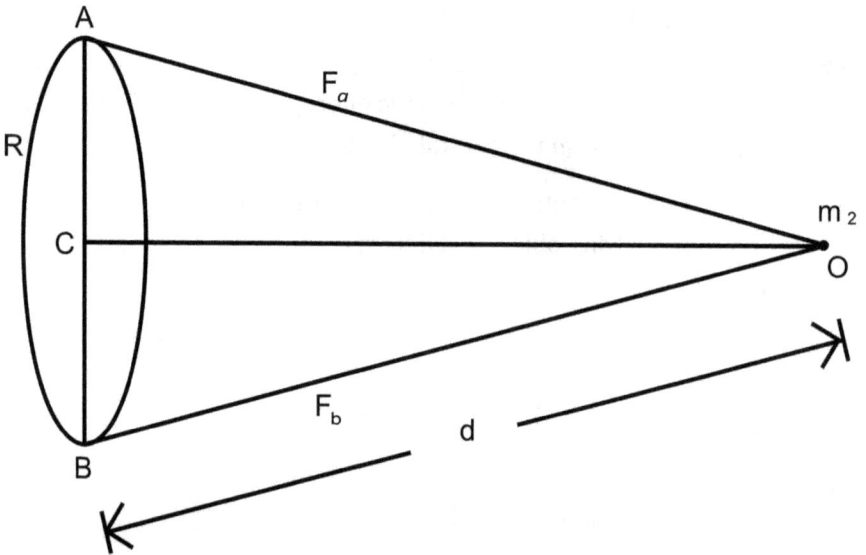

Figure 1: Gravity between a point and a ring

through its center.

As shown in Fig. 1, the ring R is a combination of point mass particles each with a mass of m. O is a point mass object with a mass of m_2. Object O is on the line going through the center of the ring. θ is an angle between any two opposite points (ex: A, B) on the ring R and the object O. d is the distance between any point on the ring and the object O, a constant value for all the points on the ring.

Intrinsic gravity between the particles A and O is F_a and the intrinsic gravity between the particles B and O is F_b. The intrinsic gravities F_a and F_b will be equal in strength.

Intrinsic Gravity , $F_a = F_b = Kmm_2/d$

Combined intrinsic gravity,

$$F_a + F_b = 2Kmm_2/d$$

If n is the total number of point mass particles on the ring then there are n/2 pairs of point mass particles on the ring.

Then the total intrinsic gravity of the ring to the point mass particle is

$$T_g = (n/2) * 2Kmm_2/d$$

$$= K(n*m)\, m_2/d$$

The product of $n*m$ is the total mass of the ring, say it as m_1.

Then

$$T_g = Km_1m_2/d$$

Resultant gravity of F_a and F_b works on the line OC towards the center of the ring.

Resultant gravity of F_a and F_b

$$= sqrt\,((F_a)^2 + (F_b)^2)$$

$$= Kmm_2 * sqrt\,((1+cos\ \theta)^2 + sin^2\theta)/d$$

Therefore the total resultant gravity of all the particles on the ring and the point mass particle is

$$R_g = (n/2)*Kmm_2 * sqrt((1+cos\theta)^2 + sin^2\theta)/d$$

$$= K(n*m)m_2 * sqrt((1+cos\theta)^2 + sin^2\theta)/(2*d)$$

$$= Km_1m_2 * sqrt((1+cos\theta)^2 + sin^2\theta)/2*d \qquad \text{--- Eq. 1.}$$

When $\theta = 0^0$, i.e, when the point mass is far away from the ring, the resultant gravity will be equal to

$$R_g = (Km_1m_2*2) / (2*d)$$

$$= Km_1m_2/d = T_g$$

When the objects are far away from each other, they both look as point mass particles to each other; therefore the equation for the resultant gravity will be equal to the total intrinsic gravity.

When the angle, θ is less, the displacement or resultant gravity will be more. Therefore, the point mass object accelerates more rapidly towards the ring. When θ approaches 180°, the displacement gravity will decrease. Therefore, in the vicinity of the ring acceleration will decrease.

When the point mass is at the center of the ring, the distance between the ring and the point mass will be equal to the radius of the ring. The angle between any two points on the ring and the center point will be equal to 180°. Applying these values in the Eq. 1, we will get the value for the resultant gravity as

Resultant gravity,

$$Rg = Km_1m_2 *sqrt((1+cos\theta)^2 +sin^2\theta) / (2*d)$$

$$= Km_1m_2 * 0 / (2*r) = 0$$

where r is the radius of the ring.

When the displacement or resultant gravity is equal to zero then the differential gravity will be equal to the whole of the intrinsic gravity.

$$D_g = T_g + R_g$$

$$= T_g + 0 = T_g$$

Finding a principle for the gravity between a ring and a point mass particle away from it is very important because many of the scenarios can be expressed using this ring and point mass combination.

A shell can be represented as a set of thin concentric rings. Gravity between a point and the shell can be represented as the sum of all the gravity between all the rings and the point mass particle.

Karunakar Marasakatla

A sphere can be represented as a set of concentric thin shells. Ring and the point mass principle can be used to calculate the gravity between each shell and the point. Gravity between the sphere and the point mass will be the sum of all the gravity of all the shells and the point mass particle.

As observed in the derivation of the gravity between a ring and a point mass, there are two forces acting on the point at any point of time. They are the resultant gravity and the differential gravity. As the size of the ring increases, the strength of the differential gravity will also increase. When the ring is also converted to a point mass, then there wouldn't be any differential gravity at all. It will all be just the resultant gravity.

The gravity between a shell and a point mass object within the shell is a non-zero value except when the point is at the center of the shell, in that case the differential gravity comes into the picture.

Chapter 13:

Gravity around different objects

Gravity between any two objects changes depending upon the mass, distance, size and shape of the objects. The strength of the gravity will also change depending upon the orientation of the objects to each other.

Gravity around a non-sphere object will not be same at equal distances because the shape of the object will change when viewed from different sides. A cone looks like a cone when viewed from a side and the same cone looks like a circle when viewed from the base of the cone. A ring looks like a circle from above or below the plane of the ring and at the same time, the same ring looks like a simple line when viewed from the plane of the ring. Rest of the chapter will explore the gravity around different objects.

Gravity around a Ring of Mass

As shown in Fig.1, if we imagine a shell *S* around the ring *R* with the radius and the center of the shell same as the ring, the gravity will be stronger at the intersection of the surface of the shell and the plane of the ring. Gravity will decrease as we go away from that plane towards the line going through the center of the ring. Gravity will be weak at the point where the line *AB* goes through the shell.

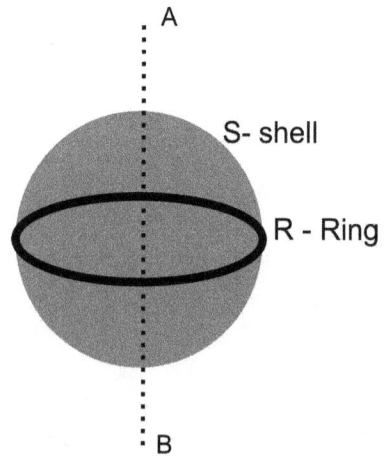

Figure 1: Gravity around a ring

In this example, gravity was compared around the ring at equal distance from the center of the ring. Surface of the shell around the ring is same distance from the center of the ring.

Gravity around a circular disk of mass will also be similar to the gravity around a ring of mass. Gravity around a disk will be more on the plane of the disk and will be weak above or below the disk.

Gravity around a Sphere

Because sphere doesn't have any orientation, the gravity around a sphere, at any specific distance from the center, will be same. If we draw a shell S, around the sphere A with a radius more than the radius of the sphere, gravity of the sphere A on any point on the shell S, will be equal as shown in Fig. 2.

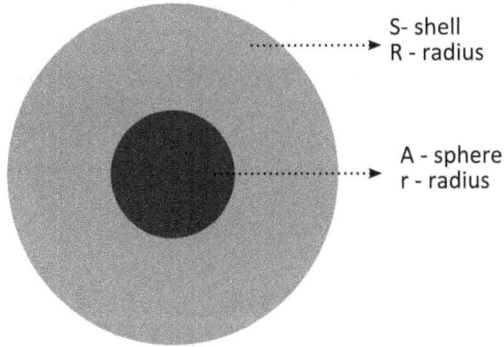

S- shell
R - radius

A - sphere
r - radius

Figure 2: Gravity around a sphere

Gravity around a Cone

Cone is not a symmetric structure from most of the cross sectional views. Therefore, it looks different from different angles outside of the cone. Gravity around a non-spherical object will change with the orientation of the object.

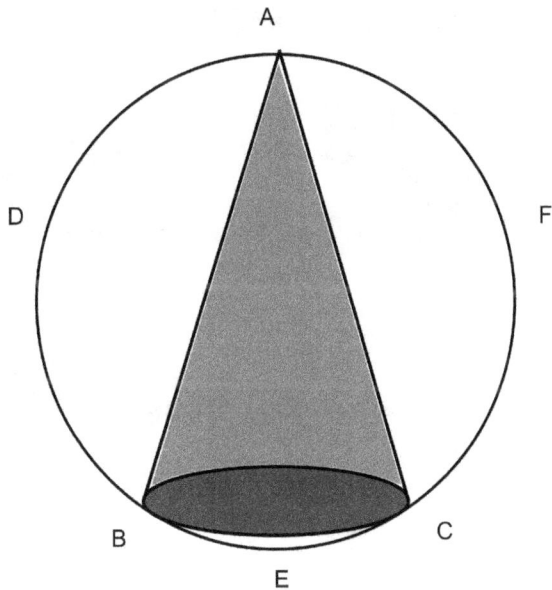

Figure 3: Gravity around a cone

Let's assume a spherical shell around the cone as shown in Fig 3.

Gravity From A New Angle

Maximum gravity will be near the tip of the cone, where the line going through the tip intersects on the shell at *A*, because, this is the only place where entire mass of the cone appears in a small angle. It will be stronger even around the rim of the base because the angle will be small all around the circle of the base, ex. at points B and C. Gravity will be weak on the shell where the lines going through the center of the sides of the cone intersects the shell, like the points *D*, *E* and *F*.

If the radius of the base is greater than the height of the cone, then the gravity will be more on the rim of the shell than at the tip of the cone.

Gravity around a Cube

For a cube, gravity will be more at all the corners of the cube, just like the tip of the cone, because it has a less angle compared to the rest of the cube as shown in Fig 4 (a). As we go towards the center of the sides,

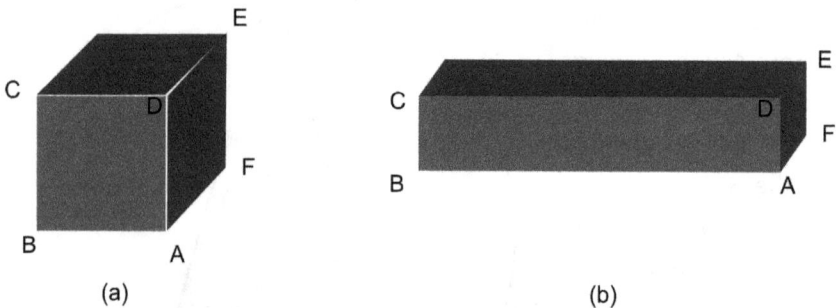

Figure 4: Gravity around a cube and rectangular bar

the gravity will decrease. Gravity will be minimum at the center of each side. For example, gravity at the center of a side *ABCD* will be minimum when compared to the gravity at all other places (except at the center of other sides) on the surface of the cube.

Gravity around a Rectangular bar

For a rectangular bar, the gravity will be more on the corners just like the cube. If the length of the rectangle is grater than the width and the height of the bar then the gravity will be more on the base and top of the bar because the angle, each of the particle at the base or top of the bar makes with any two particles of the bar will be less as shown in Fig 4 (b). Gravity will be weak at the center of the side rectangle *ABCD*.

Gravity within objects

As the gravity changes around an object, it will also change within an object. In general, the gravity within an object will decrease because the differential gravity will increase inside the object. At times, like at the center of a sphere or ring, the resultant gravity will even becomes zero as seen earlier.

Chapter 14:

Gravity is the Strong Nuclear Force

There are four known forces in the nature. They are electromagnetic force, strong and weak nuclear forces and the force of gravity. Strong and weak forces were believed to be working only at the subatomic level. Gravity was described as strong force only at greater distances. Electromagnetic force was described as working at all the places.

Electromagnetic and the strong and weak nuclear forces claimed to be in relation to each other. The only force at odds with these forces was the gravity. Combining the gravity with other forces or describing the other three forces in terms of gravity was became one of the foremost challenges of not only the physics but the entire science. Because of the importance of the unification of the forces, it was termed as the holy grail of physics.

Now we know for sure that the weight of any object at a any given location is not a fixed amount. If the object shrinks in size it weighs more and if it expands, it weighs less. If an object shrinks, it also gains more gravitational self energy because now the particles in the object are closer to each other than earlier. If the object expands, it releases the energy and contains less gravitational self energy.

The resultant gravitational force between two iron spheres of one kilogram each separated at a distance of one meter is very less. If we compress the two spheres to a point size objects, the weight of these objects will increase many fold. At the same time, the gravitational force between them will also increase many fold. Now if we decrease the distance between the objects to subatomic level, the gravitational force between them will increase even more. At this scale, the gravity between the objects will be a very strong force.

Gravity was assumed to be weak because a small magnet could able to deflect an iron bar from the gravitational pull of the entire earth. If the entire earth turned out to be a point size object located at the center of the present earth, the gravity between that earth and the iron bar at same distance as earlier would be much greater than when the earth was bigger in size.

The only two forces assumed to be working at the subatomic level are the strong and weak nuclear forces. Because of its apparent weak nature, gravity was assumed to be not working at this level. Now gravity turned out to be a very strong force between point mass objects at greater distances as well as at the subatomic level. At present, there is no other strong unknown force works at the subatomic level other than the strong nuclear force.

Therefore, all the scenarios explored above leads us to conclude that the gravity at the subatomic level is nothing but the strong nuclear force itself. This conclusion resolves many outstanding issues in physics.

There is no satisfactory explanation for the strong nuclear force not to work beyond the subatomic level. The mysterious disappearance of the strong nuclear force beyond the subatomic level can be resolved by replacing it with gravity. Gravity works at any distances including at subatomic level.

By replacing the strong nuclear force with the gravity, we are not only unifying all the forces but also removing the dilemma of gravity not working at subatomic level as well as the dilemma of strong nuclear force not working at macro level.

With this unification, the enigma of source of energy at the subatomic level will also be resolved. The energy at the subatomic level is stored in the form of gravity. It depends on the distance between the sub atomic particles.

If the protons and the neutrons are tightly bound in the nucleus, it will have more gravitational self energy and weighs more. If a tightly bound nucleus converts into loosely bound nucleus, it releases the internal energy and weighs less.

The new theory not only achieves the grand unification of all the physical forces [11] in the nature but also explains most of the observed physical phenomena in a more meaningful way.

With this unification, another dilemma of the gravity will also be removed. Because the strong nuclear force is the gravity, the nature of the strong nuclear force can be attributed to the gravity. As the Newton observed, the gravity might be working as an "Action at a distance" just like the subatomic strong nuclear force.

Chapter 15

E = mc^2, what does it represents?

Even after invalidation of many principles, one of the widely known modern equations, $E = mc^2$, appears to be valid to its core because of its apparent ability to predict the energy transformations in the nucleus.

As we concluded earlier, binding force or the strong nuclear force in the nucleus is nothing but the gravity itself at work at subatomic level.

When protons and neutrons in a nucleus are tightly bound together then that nucleus weighs more. The tightly bound nucleus will have more gravitational self energy due to its compact nature. When the nucleus split into less tightly coupled nuclei compared to the parent nucleus then the resultant nuclei will have less gravitational self energy and the difference gets released as the energy.

Even though the object has same mass, if it is compact, it will have more internal energy. If the object is large with the same amount of mass, it will have less internal energy. So the energy within the object depends upon the amount of matter inside the object as well as the size of the object.

$E = mc^2$ denotes that if an object has more mass then it will have more energy as well. According to this equation, energy within an object doesn't depend upon the size of the object which turned out to be not true.

If we consider the mass as a comparative gravity as we are using it today, then the object will have more energy if it exerts more gravity. The size of the object doesn't matter in this scenario. If we consider the mass of an object as the matter within the object then between any two objects with same amount of matter, the smaller object will have more gravitational self energy.

$E = mc^2$ holds good when the object shrinks to a smaller size. When an object becomes small, it will exert more gravity, ultimately measuring more mass according to the method of comparative gravity as the measure for the mass. More mass in the object, more energy it contains. If it expands, it measures less amount in mass; as a result it contains less energy. This analogy is true only for the method of comparative gravity as the measure of the mass. But the true representation of the mass is the amount of matter inside the object.

More or less the special relativity is a bit in conjunction with this new theory. In contrast, the general relativity, the later contribution of Albert Einstein is totally contradicts the special relativity. The core of the general relativity says that the object's gravity doesn't depend on its size. As explained earlier in the new theory, gravity does vary with the size of the objects.

The objects total energy at an instance is its gravitational self energy. If we derive an equation for the gravitational self energy of an object using the new principle of gravity and the new definition of mass then that will replace the need for the equation $E = mc^2$.

The equation, $E = mc^2$ tries to measure the entire energy within an object using its mass. Energy within an object is not a fixed amount. The energy within an object can vary from small amount to enormous amount depending on the size of the object.

The energy involved in chemical and nuclear reactions is the energy within the particles, atoms and molecules stored at a particular size of the individual objects. Then how much energy an object actually contains? There is no limit for the energy within an object. As the object gets compressed, it gains more and more energy. It doesn't depend on how many basic particles are there in the object. Any object can't sustain the minimum possible size for ever. At one point, its own forces disintegrate the object. The most compact form of the object before it disintegrates is its maximum gravitational self energy between all of its matter.

Karunakar Marasakatla

The equation $E = mc^2$ along with the shell theorem tried to capture all the characteristics of an object into a unified form.

The following are the main concepts of these principles and their validity.

- **Object contains same amount of matter even if it changes the size.** It is true but unfortunately the matter was equated with the mass of an object and the same mass was measured using the gravity of the earth which depends on the size of the objects.
- **Mass of an object is a fixed amount.** Not true when it is measured as a comparative gravity. That's why the mass has been given a new definition (volume * Mass index of the material) which makes the mass proportionate to the matter inside the object. As long as the volume of the object remains same regardless of its shape, the amount of mass inside the object will be same.
- **Gravity between two objects doesn't depend on the size of the objects.** Not true. It does depend upon the size of the object.
- **Energy within an object is a fixed amount.** Not true. As the object becomes small with same amount of matter, the net self energy of the object increases and when the object expands, the net self energy decreases.

Size, an important factor of the object was ignored in defining these principles. When we include the size of the object into the picture, all other characteristics of the object becomes clear.

Lets consider two objects made of same amount of matter, one is 1 Kg iron sphere and the other is a thin wire. The wire is stretched from end to end. According to the mass-energy equation, both of these objects contains same amount of energy. In reality, the iron sphere contains more gravitational self energy than the thin wire spread from end to end.

Mass deficit

As long as the basic particles inside an object remain same, the mass of the object should also be same because the mass is a measure for the matter inside the object. In that case, the scenario of mass deficit will never arise for any given object. Then what is the mass deficit being referenced in regard to $E = mc^2$ in the nucleus of an atom?

Mass deficit in a nucleus is the difference between the combined weight of all the protons and the neutrons or the nucleons inside an atom to the weight of the nucleus they are in. Nucleus always weighs less than the combined weight of the nucleons.

The loss in weight for the nucleus is nothing but the loss of gravity between the nucleons and the earth when they are occupied more space. If all the nucleons were packed together in a size of a single nucleon, then the weight of all the nucleons will be equal to the combined weight of all of the nucleons. When the nucleons expand and occupy more space in the nucleus, the difference in gravitational self energy will be released as energy.

The loss in weight due to gravity of the earth has no relevance to the energy within the object.

Two objects with different shapes made out of same amount of matter might weigh same measure in gravity. Due to their difference shape, the binding energy within these objects will be different.

The flaw in shell theorem, change in gravity with the size of an object, is more evident in the mass deficit than any where else. Deficit in the mass of a nucleus when it is measured as the comparative gravity is natural if that object expands in size.

Chapter 16:

K & G - The Case of Two Constants

The value of the G, the universal gravitational constant is precise only to few decimal places. It even varies between each attempt and never exactly same between the derivations of two different groups [12, 13]. The most commonly accepted current value of the G is 6.123333 Newtons.meter^2/Kg^2.

None of the two experiments attempted at deriving the value of the G used same set of parameters. The size of spheres used in the experiments varied between each experiment. With the invalidation of the shell theorem, it is now evident that the gravity differs between same mass objects with different sizes. The variance in the value of G is natural for these experiments because they were conducted at different places on the earth with different sizes of spheres and even with a different distance between them. Gravity between objects varies in all of these scenarios.

Because of the invalidation of the Newtonian principle, the value of the G stands for nothing in the physical sense. No wonder it was varying uncontrollably in a wide range.

At one point of time, even the value of the G was proposed as decreasing with the time from the beginning of the universe. It was thought that in the beginning the objects were having more gravity between them and it gradually leaked into other universes.

It is still a wonder in scientific community that an object of the size of earth exerting so small amount of force on the objects on its surface. Gravity of the earth pulls the object on its surface in as wide angle as 180^0. Even though the total gravity of the earth is strong, the resultant force of the gravity is weak. Therefore the object on the surface is weighing very less when compared to the size of the earth.

Unlike the universal gravitational constant, the value of K, the new constant of intrinsic gravity, doesn't vary with the size or shape of the objects because it was defined for the smallest possible object, the point mass object.

The equation for the new constant will be as follows.

$$K = Fd / m_1m_2,$$

F = The energy consumed between the objects in a second (force is power).

d = Distance between the objects

m_1, m_2 are the masses of the two point mass objects.

Whatever the standards we use to derive the value of the K, it will remain a constant.

The numerical value of K can be defined in the MKS system as the energy generated between two one kilogram point mass objects separated at a distance of one meter in a period of one second. The mass of the objects will change at the end of the second by an amount dependent on the energy released.

Because we can't create and measure point mass objects, a new principle was defined for non-point mass objects.

$$C = Fd/m_1m_2$$

F = The energy consumed between the objects in a second (force is power).

d = Distance between the objects

m_1, m_2 are the standard masses of two standard volume objects.

The value of C depends on how precisely we define and measure the mass of an object. As we approximate the value of the standard volume towards the point mass, the value of the C will tend towards the K.

Karunakar Marasakatla

Decreasing mass in a Standard Kilogram

Earths gravity is decreasing gradually because of the loss of energy due to the gravitational pull on the moon.

Objects on the surface of the earth also loose their internal strength because of the earth's gravity. An object kept on the surface of the earth is not part of the earth. Earths gravity will continuously strain the object of its internal energy. An object kept on the surface of the earth will gradually weigh less.

If the local gravity is strong then the object will loose its internal energy even quicker than the normal.

An object also radiates energy at the nucleus. Any object that radiates energy becomes lighter in comparative gravity.

The standard kilogram bar kept at the International Bureau of Weights and Measures might be losing its weight [6] because of its self radiation of energy or due to the degradation of the material because of the continuous pull of the local gravity on the bar or both. As an object decreases its weight, the size of the object will increase. The volume of the bar will increase if it is losing the weight.

Part III: Confirmations, Predictions and Challenges

(Applications of the new theory of Gravity)

Each new theory has to incorporate all the phenomena that have been successfully explained by the existing theories. And also, it has to explain the phenomena that has been observed but not explained by the existing theories, the so called anomalies. Explanation of the observed phenomena will be the confirmation of the new theory.

The new theory should also be capable of extending our vision of understanding the nature by predicting the possibilities not yet observed. Later experiments should confirm these predictions.

Each new theory has challenges too. The predictions of the theory should be confirmed at some point of time. And also more specifically, the theory should be able to explain new observations in the nature without deviating from the basic concept of the theory.

Even though there were some descriptions of the observed phenomena in the earlier sections of this book, this section is particularly devoted to explain most of the other observed phenomena along with all the gravitational anomalies using the new theory.

Chapter 17:

Singularity and the Black Holes

When all the matter of an object compressed to a point size then it is called a singularity. All the prevailing theories wouldn't differentiate between an object turning into a point size or expanding to even a bigger size as shown in Fig. 1, because both the objects were having same amount of material or in other words, mass.

Object *B* is the expanded form of the object *A* and object *C* is the point mass form of the object *A*. All the three shapes of the object *A* in the Fig. 1 were treated as same for measuring the gravity because of the shell theorem.

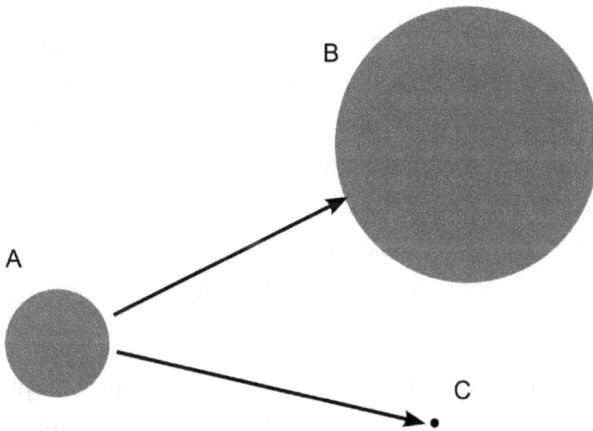

According to this new theory, even though all the three forms of the object has same amount of matter, all three forms of the object measure differently for the standard mass. As the object becomes smaller, its standard mass will increase and at the same time its gravitational self energy will also increase because the atoms and the particles within the object comes closure to each other.

Figure 1: Singularity: Mass and size of an object

As the matter compresses, at some point of time, the gravitational self energy of the object becomes enormous and the particles can't sustain such energies for a long period of time. It eventually disintegrates by releasing the energy and becomes normal matter.

Black holes were described in the standard theory as singularities or point mass objects. It was calculated using the current theories that if the earth compressed to a size of few centimeters wide, the light wouldn't even escape on its surface because of the strong gravity on the surface of the compressed earth. If earth ever becomes that small, the gravity of the compressed earth will be so strong that it will trap the light even away from the surface of the compressed earth.

If there is ever a black hole and if it persists for a long period of time, it devours all the material around it, eventually consuming entire universe. But the universe doesn't seem to be ruled by the objects called black holes. They can't form from the stars and even wouldn't exist forever devouring the surrounding objects.

The objects exerting more gravitational force than the neutron stars might be neutron stars with more mass or they could even be made of more compact material than the neutron stars. But it can't be a singularity. Strong gravity or the high energy emission we observe from these objects is because of the energy loss due to the material revolving around that object. High energy emission of the galactic center is the loss of its gravitational self energy due to the revolution of the stars that are orbiting in the entire galaxy.

As long as there is material revolving around the object, the matter will never collapses into a point. The matter around the object keeps pulling the object at the center from all sides. This pull from the revolving matter prevents the central object from collapsing. If there is no material around the object, it will rotate fast to prevent from collapsing.

Curvature of space

Around bigger objects, like the Earth and the Jupiter, the differential force will be more than the differential force around the same objects if they would have been point mass objects. The differential force of these objects makes the objects around it lighter in weight and enables them to revolve around the body easily. The escape velocity of a revolving object will decrease if the size of the central object increases, because of the decrease in the resultant gravity of the central object as it increases. An approaching object tends to go around these objects.

The tendency of the falling objects to go around a bigger object might appear like the space around it is curved. It simply an effect of the differential force generated by the wider distribution of the mass.

If a bigger object like the Jupiter turns to a point mass object, then the differential force around that object will be zero. The total gravity acts towards the object, means the total gravity is the resultant gravity. In this case the escape velocity around that object will more than when it was a bigger object. An approaching object will fall straight into the object. Here the falling object might disintegrate due to the tidal forces, a difference in strength of gravity between different parts of the falling object.

Chapter 18:

Galactic Rotation Curve and Dark Matter

In the 1960's, it was observed that the stars on the outskirts of a galaxy were revolving around the center of the galaxy faster than the current theories could predict. Naturally, gravity decreases with the distance. The stars in the outer orbits should be bound to the galactic center with less gravity.

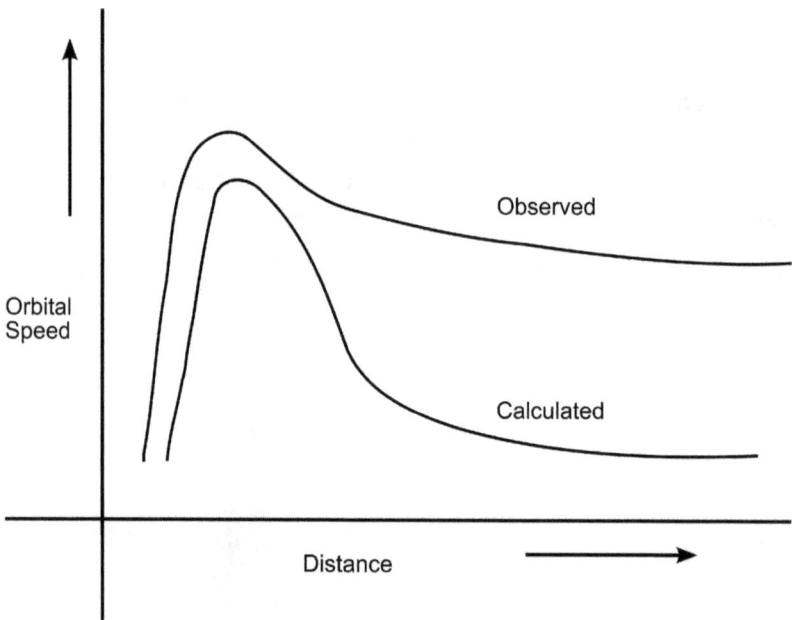

Figure 1: Galactic rotation curves

It was reasoned that if the stars to remain in the orbit around the galactic center, their orbital speed has to be less than what have been observed. If they move faster, they will fly away from the grip of the galaxy. Because the stars were bound to their galaxies for a long period of time even with a higher orbital speed, something like a strong force should be working in pulling the stars towards the center of the galaxy or make them bound within the galaxy.

When Vera Rubin [14] plotted the observed speed of the stars and their distance from the galactic center on a graph against what the current theory predicts, both the values match for a distance, after that they differ drastically as shown in Fig 1.

The data displayed in the graph is not specific to any galaxy; it was a general description of the galactic properties.

The strange force that is holding the stars to their galaxies remained a mystery. To resolve the anomaly, scientists proposed the presence of dark matter [1,2,3] in the galaxy. Dark matter particles were presumably made of not yet discovered exotic particles which were heavier than the particles of the visible matter. They were also weakly interacting with the visible matter. Hence they were termed as Weakly Interacting Massive Particles (WIMPS).

The dark matter theory also proposes that more than 90 percent of the matter in the universe is the dark matter. The dark matter in our galaxy, for that matter in all other galaxies where the stars at the outskirts were orbiting faster than normal, is assumed to be encompassing the entire galaxy in a shell.

Instead of inviting the dark matter with exotic nature, is there any natural cause for the observed phenomenon?

It is widely agreed that the object at the galactic center is more massive than any other object in the galaxy. Certainly the stars close to the galactic center are orbiting around it due to the strong gravitational pull of the galactic center.

Is the galactic center a singularity (a point mass)? As discussed earlier, singularity can't exist for a long period of time. Therefore, it is possible that the galactic center is a densely compressed spherical object with a radius possibly equal to the distance of the event horizon from the center of the galactic center.

The stars closer to the galactic center will have less resultant gravity of the galactic center due to its bigger size like the sun exerting more differential and less resultant gravity on the planet Mercury. As the orbital distance increases, the galactic center looks like a point. At an orbit at the outskirts of the galaxy, the massive galactic center and the mass inside that orbit looks like a point mass therefore will have more displacement or resultant gravity than the differential gravity. Because the gravitational pull of the galactic center and the mass inside the orbit is more on the stars at the outskirts, they are orbiting faster to balance the gravity.

According to the new theory, gravity between the galactic center and a star at the outskirts of the galaxy will be stronger than the gravity calculated using the standard theory because of the invalidation of the inverse square law of Newtonian gravity.

If the dark matter surrounds the galaxy in a shell, the stars will spiral down towards the galactic center.

Therefore there is no need to invoke the dark matter theory to explain the orbital speeds of the stars on the outskirts of a galaxy. It is a natural phenomenon of the gravity caused by the point mass object on distant objects.

It appears that the galactic center does really hold enough mass to exert gravity on the solar system to hold it in an orbit. All these days we were of the opinion that there was not enough mass inside the orbit of the solar system to gravitationally bind the solar system to the galactic center.

Is the search for the dark matter over?

Now we know that the huge amount of exotic dark matter is not required to explain the anomalous galactic movements then what else the underground lab in Soudan, Minnesota and the Large Hadron Collider in Europe could able to find?

Karunakar Marasakatla

Still, these two facilities might find new matter and extend our understanding of the nature of the matter but the amount of the new matter will definitely not exceed the amount we already see in the universe.

Chapter 19:

Pioneer Anomaly

According to the new theory, the gravity on other planets and their satellites will drastically change to the one on earth because of differences in size and density between the objects and the earth. Shape of the objects doesn't matter because all are being in spherical shape. Even the gravity on our moon will be different than what we have calculated according to the present theories. Yet, we were successfully able to launch numerous probes onto these objects without any difficulty. Are we?

If the current theories are wrong, there should be at least one abnormal behavior of these numerous satellites launched into the space should have been observed.

Apparently not one but many abnormal behaviors of the deep space probes were observed which were not explained by the current theories. Among them, a prominent one is known as the pioneer anomaly.

Pioneer 10 and Pioneer 11 were two space probes that were launched into space in 1970s by NASA to study the outer objects in our solar system. After the completion of the planned mission, the probes were directed out into the deep space away from the sun. Luckily, the probes were alive for a long period of time and were able to send information to the control center. By studying the data sent by the deep space probes, NASA scientists [15] found that the probes were a bit slowing down and veering off the destined path. After eliminating all the possibilities, it was observed that it is an anomaly which can't be explained by existing theories and was named after the probes as pioneer anomaly.

According to this new theory, it is natural for the probes to slow down at those distances from the sun. Even though there is lot of material in the sun, due to its enormous size, the resultant gravity will be weak in the vicinity of the sun, for example from its surface to the orbit of the Mars where it appears bigger than a point in the sky. As we go away from the Mars, the sun appears smaller and the displacement or resultant gravity will increase.

Deep in the space where the probes were shown abnormal behavior, sun looks like a point mass. Resultant gravity of the sun will be stronger in these outer regions of the solar system. All of the sun's gravity works on a line joining the object and the point mass sun; therefore it was able to slow down the pioneer probes a bit more than what was expected.

Pioneer anomaly was not limited to the two pioneer probes. It was observed with many other probes those extended their journey deep into the space.

Not only the sun, even the planets will also exert this additional force at a distance where they will look like a point. We will find this same additional force in the vicinity of the earth where it starts to look like a point. Even this additional force of the earth appears to be observed in one of the probes.

Chapter 20:

Differentiation of the Earth and Formation of the Moon

The process of differentiation in any object is an amazing phenomenon. It is interlinked with many of the processes related to that object. Because of the standard definition of mass and the shell theorem, it was not given much importance in science.

Differentiation of the Earth

Earth was uniform in density when it was coalesced from the dust particles in the beginning as shown in Fig. 1 (a). Gravity around the initial earth will be just like

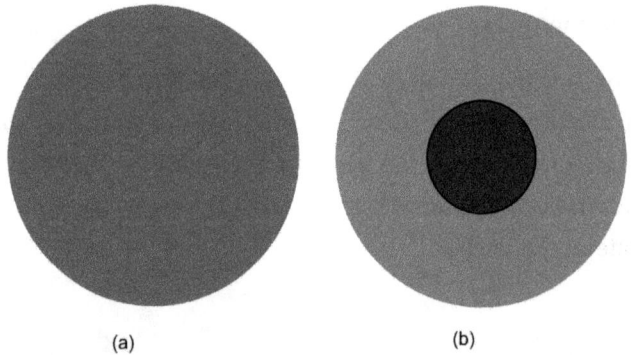

(a)   (b)

Figure 1: Gravity around a uniform and differentiated earth

around any sphere object. When earth differentiated, all the denser material migrated towards the center of the earth and formed as a core and the rest of the body dominated with lighter material as mantle as shown in Fig 1 (b).

Now, the gravity exerted by the material of the core will be stronger than the material of the uniform earth. More over the core occupies less space inside the earth. Angle of projection of any object on the surface of the earth to the core will be small compared to the whole earth.

Therefore the combined resultant gravity exerted by the differentiated earth will be more than the differential gravity on any object on the surface of the earth. In the case of uniform earth, the differential gravity will be more than the resultant gravity on the object on the surface of the earth.

In essence, an object on the uniform earth would have weighed less than the present differentiated earth due to less resultant gravity on the uniform earth.

Formation of the Moon

Any collapsed object from the dust particles will have at least a minimal amount of angular momentum due to the collision between the collapsed particles. Let's assume that the initial earth was rotating lot slower than the current speed.

We all agree that a slowly rotating skater with the arms stretched out wide will rotate even faster when the arms are pulled towards the body.

What if the entire body of the skater is concealed inside a big balloon? We can't see the movement of the hands or the body of the skater inside the balloon. We can only see the rotation of the balloon.

When a rotating skater with arms stretched wide inside the balloon, pulls back the arms towards the body, the balloon starts to rotate faster than earlier. Physically, we wouldn't see any change in the appearance of the balloon.

Current theories assume that there wouldn't be any change in the object because there is no change in the size of the balloon. All the theories about an object revolve around the mass of the object and that mass is assumed as a constant for any given object even if the material displaces from one place to another within the object.

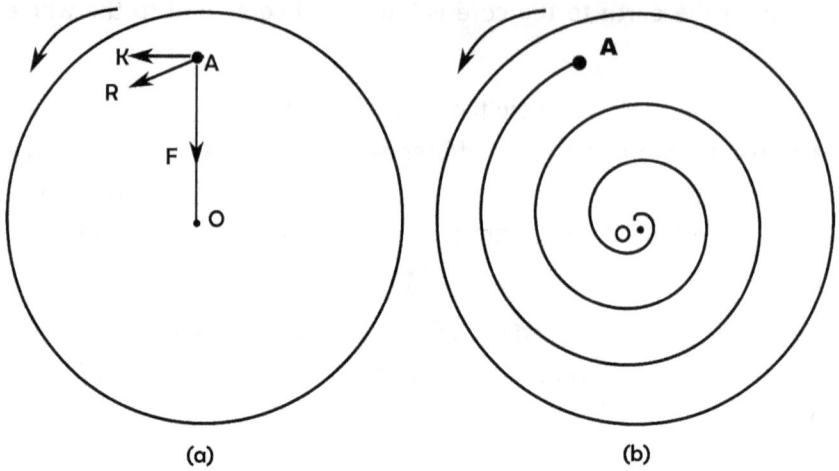

Figure 2: Process of differentiation in a planet

Current geophysical theories propose that heat is generated inside the earth in the process of differentiation. As the differentiation itself takes place when the planet is in molten state, further generation of heat in the molten planet evaporates the material in the planet. Therefore the notion of heat generation in the process of differentiation of a body is not a realistic one.

As in Fig. 2 (a), the circle is the equatorial cross sectional view of the early uniform earth with anti-clock wise rotation. O is the center of the earth and the point A is a particle of heavy material. K is the kinetic energy of the particle. F is the resultant gravity of the entire earth on the particle. R is the resultant force of forces K and F acting on the particle. The direction of the R gradually points towards the center of the earth. The direction and the magnitude of the R will provide the particle more momentum as it spirals down towards the center of the earth as shown in Fig. 2(b). As a result, the rotational speed of the earth increases.

As more and more material spiral down towards the center, the speed of rotation increases. Basically, the increase in the gravity between the particle of heavy material and the entire earth converts into the angular

Karunakar Marasakatla

momentum of the object. As each atom of a heavier element migrates towards the center of the earth due to gravity, an atom of a lighter element will be pushed towards the surface. As the material accumulates at the center of the earth, the gravity of the earth increases on the heavy material that is migrating towards the center. As the differentiation of the earth progressed, it gained more momentum and at the same time lighter elements accumulated on the surface of the earth.

It is also an observed fact that the earth contained water on its surface as early as when it started the differentiation. The presence of water on the surface solidifies the outer layer of the earth.

As the earth gained more momentum when the differentiation progressed, the material inside the earth starts to rise up around the earth in the wide ocean as the present Hawaiian volcano. Initially, the volcano starts to build at the bottom of the ocean and eventually surfaces above the ocean. Because the differentiation is a slow process, the momentum of the earth also increases slowly. The volcanoes all around the earth also grows gradually with the increase in the momentum. Once these volcanoes are above the surface of the water, they spew the lava out into the air. As the earth increased its momentum, the spewed lava revolves around the earth instead of falling back onto the earth.

All the revolving material coalesces to form the early moon. As the earth continues to eject the material, the moon continues to grow in size. If the moon is in sync with any one of the volcanoes, that particular volcano grows more than any other volcanoes because it feeds the moon with more lava than other volcanoes because of the extra pull of gravity of the moon on the volcano.

Earth-Moon system stabilizes when the earth completes the differentiation. Earth, at this stage, might have been rotating with maximum speed in its history in the presence of fully formed moon.

This simple explanation satisfies most of the observed phenomena in the earth-moon system.

- Earth got its angular momentum from the differentiation of its own body.
- Moon contains all the lighter material similar to the mantle of the earth.
- Age of the moon coincides with the period on the earth when earth underwent the differentiation.
- Moon is basically an accumulation of lava. This explains the high temperature origin of the moon.

From this, we can conclude that any celestial object that differentiates will rotate faster than normal. More over chances of that object having satellites around it are more. Mars, another terrestrial planet also rotates faster than normal. The presence of the solid core and the speed of the rotation of the planet are an indication to the same process for the formation of at least one of its moons as the earth. The other two rocky planets, Mercury and the Venus don't have much angular momentum. These two planets might have not differentiated as much as the earth and the Mars. The presence of a tallest mountain in the entire solar system on the surface of the Mars, the Mount Olympus is an indication to the process of feeding one of the moons as it grew from the material inside the Mars. There is no other possible explanation for a single volcano to grow that big on the surface of a planet.

Surprisingly, the process of formation of the moon is similar to the description of moon formation in ancient Indian texts. The texts states that the moon formed from the earth. Mount Meru was described as the conduit from where the moon was fed from the earth. It was also described that the Mount Meru was the tallest mountain on the surface of the earth. If the mountain existed, the chances of it eroding in the four billion years of geological processes are more.

Early fission theory [16] for the formation of the moon states that a piece of the earth was separated and formed as the moon when the early earth spun more rapidly. The present Pacific Ocean basin was pointed as the evidence for the material that formed the moon. Only thing was not clear at that time was how the earth itself was having that much angular momentum in the early period of its formation.

In the later version of the fission theory [17], the angular momentum issue was resolved by proposing that the differentiation of the planet resulted in the increase in the rotation of the planet. As the differentiation itself is a slow process, a chunk of piece flying away from the body at an instance is not a realistic scenario. More over, pointing the pacific basin as the evidence for the place from where the moon came from is against the more evident theory of plate tectonics.

As there is no need for the dark matter to explain the observed galactic rotation curve, the theory of the dark planet with so many unrealistic conditions [18] is also not necessary to explain the formation of the moon. There are flaws not only in the present impact theory for the formation of the moon but also in the geophysical theories for the differentiation of the earth. These all are isolated static theories. When we remove all the flaws in these theories, the differentiation of the earth and the formation of the moon becomes an interconnected dynamic system.

Any rotating irregular shaped object rotates even faster when the material falls close to the center of rotation from the irregular parts of the body. When a rotating rectangular body turns into a sphere, the new sphere object will rotate faster than the bar object.

Irregularly shaped celestial objects like the comets or asteroids might gain acceleration in the rotation when material falls from far off edges to close to the center of rotation. YORP effect, radiation from surface of the object was also observed as another cause for the acceleration of the rotation of these celestial objects.

Chapter 21:

Gravity of Point Mass Celestial Objects

There were many observations during the period of eclipse that were not explained using the standard theories, such as local gravity anomaly and the strange behavior of the pendulum.

Planets and stars appear like a point when viewed from the surface of the earth. The gravity between a celestial point mass object like a star or a planet and an object on the surface of the earth will be stronger than what was assumed according to the current theories.

Gravity between the objects on the surface of the earth and the point like celestial objects takes the intrinsic form of gravity because of the angle or the projection one object makes on the other object is zero degrees.

On the earth as a whole, the gravity might be weak between the earth and the point mass celestial objects but on the individual objects on the surface of the earth, the gravity of the celestial object might be stronger.

During the eclipse, when the cores of the earth, sun and the moon were in line, the gravity between an object on the surface of the earth in line with all three cores and the moon and the sun combined will be even stronger because the gravity exerted by the cores are stronger than the gravity exerted by the rest of the body.

Gravity anomaly on the crust [19] and the anomaly in the movement of the pendulum during the eclipse [20] might have caused by the combined gravity of the moon and the sun when they were in line.

The difference in the gravity of these celestial objects will be more visible if the objects on the surface of the earth are in lengthy cylindrical shape with less cross sectional diameter like the rod of the pendulum.

As a result of the combined gravity of the Sun and the moon, the object possibly becomes a bit lighter than normal.

Chapter 22:

Perpetual Motion

Perpetual motion, like alchemy, was part of the human endeavor for centuries to succeed. From ancient times, there were many claims and counter claims for the achievement of perpetual motion.

Perpetual motion is something similar to a continuously turning wheel on a fixed axis without any external energy in the system. It is evident that the energy can't be created or destroyed; it can only be transferred into another object or transformed into another form. Then how the wheel can continuously turn on its axis without any external energy input?

Gravity between an object and the earth depends not only the mass, size and shape of the objects but also the objects orientation towards the earth. Consider two bars of iron with same mass, size and shape. One is suspended horizontally (object A) and the other is suspended vertically (object B) from a fixed point above the surface of the earth as shown in Fig.1. The center points of the objects are parallel to the surface of the earth, in other words they are at same distance from the center of the earth.

The resultant forces a_1, a_2, a_3, ... a_n of the earth gravity on each of the particles on object A are all spread along the objects length in a wide angle. The final gravity of the earth on this object will be the resultant force of all these forces combined. Because they all are spread along a wide angle, the resultant force of all these forces will be less than the sum of all the resultant forces a_1, a_2, a_3, ... a_n. The total gravity will decrease if the length of the bar increases.

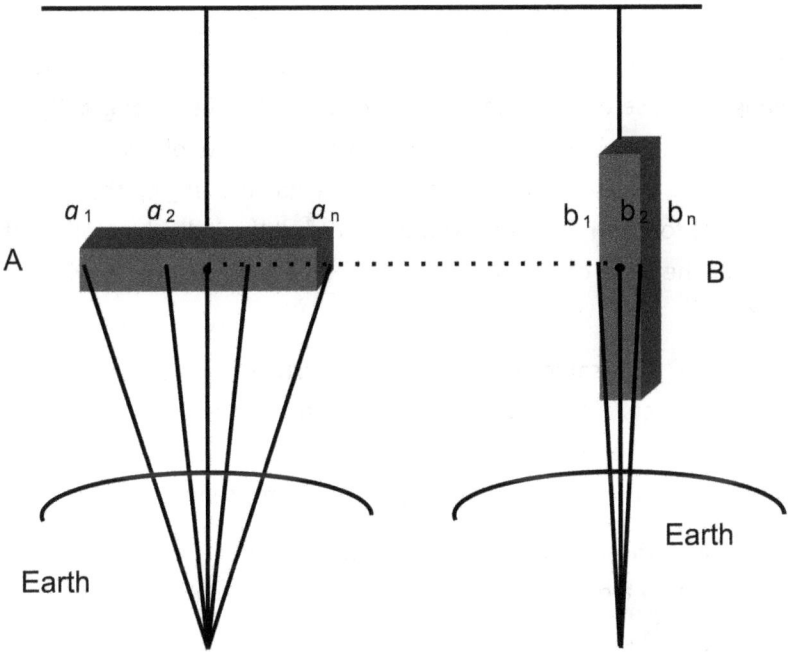

Figure 1: Difference of gravity between bars

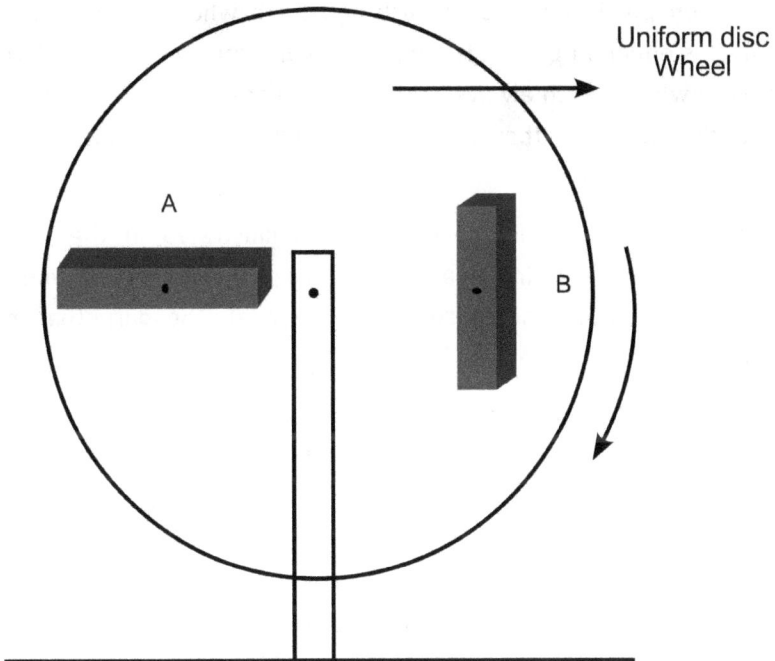

Figure 2: Unequal gravity in a wheel

Gravity From A New Angle

When we consider the object B, the resultant forces b_1, b_2, b_3, ... b_n of the earth gravity on each of the particles on object B are all grouped along the base of the object in a narrow angle. The total gravity of the earth on the object will be the resultant force of all these forces combined. Because the forces are in a narrow angle, the combined resultant force will be close to the total of all the individual forces. If the area of the base of the bar is small, then the combined gravity will be more.

Therefore, at a same location on the earth the gravity between two similar objects differs depending upon their orientation towards the earth. The difference in gravity between the objects will be significant when the density and the length of the objects are more.

If we keep these two bars on a wheel as shown in Fig. 2, the wheel rotates about a quarter of its perimeter without any external input.

Center of both the bars are from same distance from the axis. At this position, the wheel rotates clock wise about a quarter of its perimeter and then possibly oscillates back and forth when the weight of the wheel is equal on both sides and eventually comes to a halt. If we can build a wheel which always has a mass on left with a horizontal shape and the mass on the right with a vertical shape, the wheel will continuously rotate.

Surprisingly, the same mechanism was illustrated in the works of Bhaskara, an ancient Indian astronomer as perpetual motion wheel with half filled mercury in the spokes of a wheel. The perpetual motion wheel was also referred as Bhaskara wheel.

At any instance, the mercury in the spokes forms in a vertical shape on one side and the other side forms as a horizontal shape. The side with the vertical shape will have more gravity than the side with horizontal shape because the gravity will be more on a vertical object. This difference in gravity enables the wheel to rotate continuously. If properly designed, this wheel continuously rotates without any external input. Then what is the source for the energy being dispensed in this system? Earth is expending its internal gravitational self energy to keep pulling the objects towards its center.

If we drop a stone from a height up in the air, earth pulls the stone at the expense of its gravitational self energy. Earth looses a part of its gravitational self energy in the form of radiation and the stone gains the momentum.

Orbital motion of the Moon

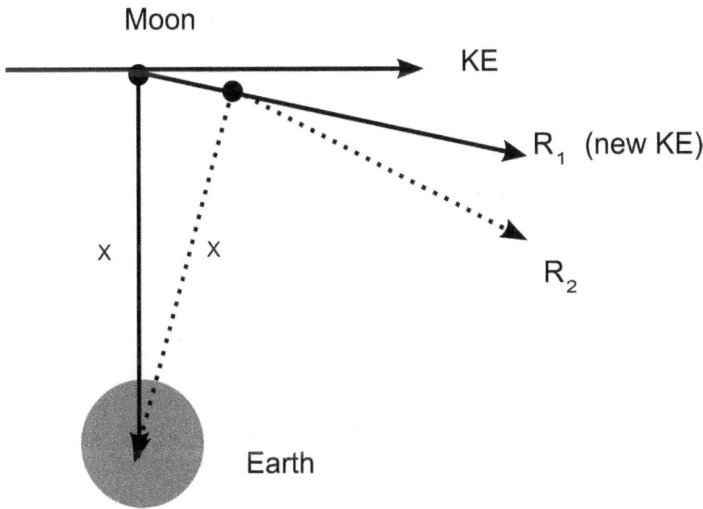

Figure 3: Orbital motion of the moon

Orbiting moon is a perfect example for the perpetual motion. Kinetic energy, *KE* of the moon keeps it in a straight line and the gravitational energy being expensed by the earth, *X* pulls the moon towards the center of the earth as shown in Fig. 3. The resultant force, *R₁* of these two forces acts in a circular path around the earth and keeps the moon in an orbit around the earth. Earth continuously looses its gravitational self energy and in the process radiates more energy in holding the moon in its orbit. The energy being radiated from the earth might be the cause of the mysterious hum observed around the earth. Not only the earth

Gravity From A New Angle

but all other planets with satellites should also radiate energy while expending the gravitational self energy to keep the satellites in the orbit.

Earth might have lost tremendous amount of its internal gravitational self energy from the time the moon orbiting around earth. In this period, because earth lost its gravitational self energy, it might have expanded to some extent, not necessarily to the extent of supporting the expanding earth theory. Gravity will also decrease on the earth from the formation of the earth to the current time due to the loss of gravitational self energy or expansion of the planet.

We use the energy to put the artificial satellites into the orbit. Once they are in orbit, they simply revolve around the earth continuously at the expense of the earth's gravitational self energy. As we send more and more satellites into the orbit, we are consuming the energy inside the earth.

Chapter 23:

Gravity Hill

There are some places on the earth where the gravity is stronger than the normal. Strength of the gravity in those places is so strong that it could even pulls the cars uphill. These locations are fittingly called as gravity hills.

Gravity hill goes against the standard theory of gravity. According to the standard theory, all the material inside the earth uniformly attracts a body which is outside of the earth. It is possible for the local gravity variations to exist because of the terrain or the material in the ground. For a strong attractive force to exist inside the earth in a particular place, the material that is causing the strong gravity has to be enormous in size as well as in mass according to the prevailing theories. The distance of the material has to be close to the surface, so that the gravity exerted by the material will be strong enough to cause the gravitational effect on the objects on the surface of the earth.

For an object to cause more gravity than the earth, the object has to be bigger than the earth. The possibility of such an object to exist inside the earth at a locality is almost nil. If the object that is causing the local gravity is small, then the current theory can't explain the strong gravity from a small object. It is a situation we can't explain by using any of the conventional methods. So the effect was outright rejected as not being caused by the gravity. Because the effect appears to be real, alternative theories were started to populate to explain the phenomenon and to support the standard theory.

One of the main alternative proposals was that the visible effect of the gravity hill is merely an optical illusion. If it is simply an optical illusion, many such attractions would have popped up around the world one each in every neighborhood.

Is it an Optical Illusion?

In many gravity hill locations, spirit level was used to demonstrate the hill as a mere horizontal position. If the cars are going uphill, a spirit level was placed at that position and if that level is in horizontal position then the hill was dismissed as a hill altogether and renamed it as a slope.

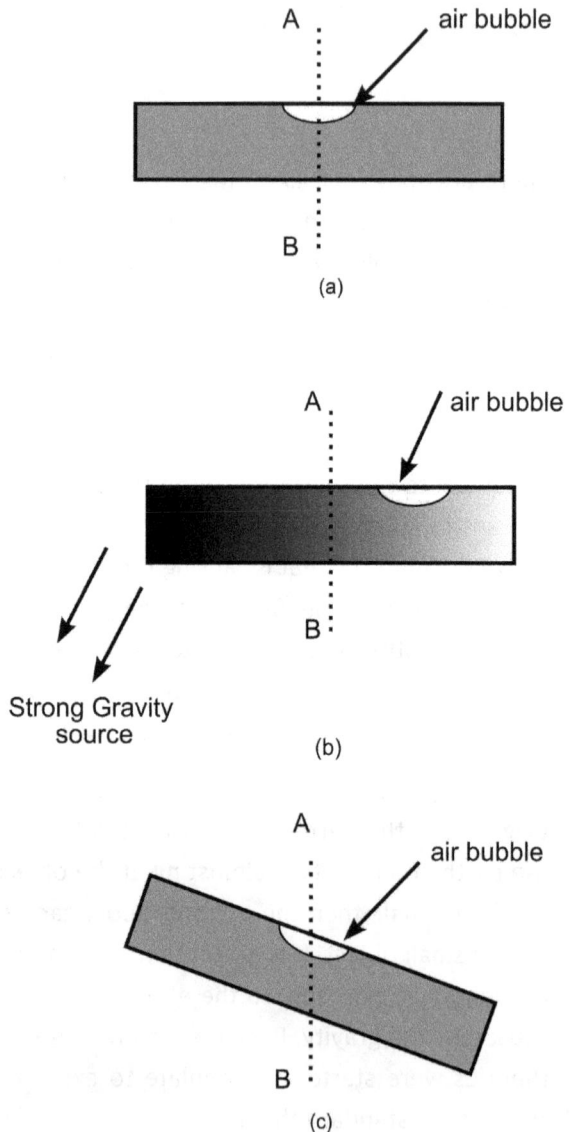

Figure 1: Effect of gravity on a spirit level

The basic principle behind the functionality of the spirit level itself is gravity. There is no other way for the spirit level to recognize the horizontal position. Air bubble in the spirit level will be in the middle as shown in Fig. 1(a) under uniform gravitational conditions. When a spirit level is placed near a strong gravity source, the spirit on the side of the strong gravity will be compressed and exerts more pressure in the tube as shown in Fig. 1(b). As a result the air bubble moves to the lighter side of the gravity. To make it horizontal, we need to lift the tube on the stronger gravity side upward as shown in Fig. 1(c). At this position the spirit level rests in horizontal position means the bubble rests at the center. This makes to believe an upward inclination as a slope, an apparent optical illusion.

There is not even a single experimental evidence for the optical illusion on the gravity hills except using the spirit level.

What is the Gravity Hill?

To gravitationally attract or pull a car up hill, we don't need an object with enormous size and mass. As shown in the Fig. 2, an inclined and denser deposit of layer such as banded iron formation surrounded by a less dense material is sufficient to exert more force on the object on the slope. Gravity will be stronger where the layer is pointing to on the surface. It will gradually decrease when an object moves away from that point on both sides. Gravity around a thick layer works as a gravity around a disk.

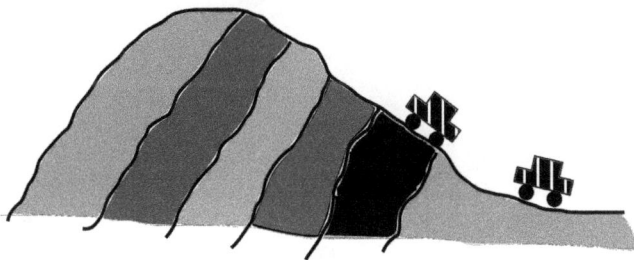

Figure 2: Gravity Hill

The difference in gravity around a gravity hill can be easily tested using a spring based weight measurement device. An iron bar weighing 1 Kg in normal gravity will be more when measured at the gravity hill. The stronger force at the gravity hill will pull the spring even further causing the same iron bar to weigh more.

Gravity From A New Angle

The difference in gravity can also be tested using the balance weight measurement system. In a normal gravity, 1 Kg of iron bar and the same amount of loose cotton will balance each other. When we balance the same materials on the gravity hill, the iron bar side of the balance will weigh more than the cotton. Iron bar is more compact compared to same weight of the loose cotton; therefore the gravity between the iron bar and the denser layer will be more compared to the layer and the cotton.

Gravity hill will be a best place to exemplify the effect of the gravity on a perpetual or gravitational motion machine. Perpetual motion machines take advantage of minute variations of gravity on both sides of a wheel. If the machine is close to a stronger source of gravity, the difference in gravity will be multiplied making the strength on one is more greater than the other. This ultimately gives enough force to rotate the wheel.

A gravity hill which can pull a car can definitely generate enough energy to give light to its surroundings.

Chapter 24:

What is a Chemical Reaction?

Chemical reactions, at the basic level, appear to be transfer of electrons from one element to another. Yet, in some reactions energy is released and in other reactions, energy is consumed. Where is this energy stored and how it is being released? If electrons are the only ones being transferred from one element to another element, are these electrons are the medium and storage of the energy involved in the reactions?

Before answering that question, let's explore about what happens when we take out or add electrons to any atom. When a neutral atom acquires or releases electrons, it is called ion. Positive ion is the one which releases the electrons and the negative ion is the one that gains the electrons.

An electron escapes from the atom when energy applied to it. Energy will be used to break the bond between the nucleus and the electron. This is similar to the satellite orbiting the earth, which escapes from the gravity of the earth when its speed is increased. Energy required for the electron will increase as it orbits close to the nucleus because of the gravitational attraction of the nucleus and the attraction of the positively charged protons in the nucleus. In any scenario, for the gravity to be strong, the central object has to be very small compared to the radius of the orbiting object. Otherwise the total gravity will be split between the resultant and differential gravities. Nucleus of the atom is very compact compared to the radius of the atom; therefore the gravitational pull will increase even to the first orbit of the electron.

Nucleus of the atom will collapse or compress into a smaller size and gains more gravitational self energy when the atom loses an electron because the nucleus is now under less stress from the pull of the orbiting electrons.

Energy required to release an electron from a gaseous neutral atom is called ionization energy. Ionization energy will increase as we go closer to the nucleus because the electrons are bound more tightly with the nucleus when they are closer to it. It requires more energy to break the bond between the nucleus and the electrons closer to it. The initial energy required is called the first ionization energy and the later are called second and third ionization energies depending upon the level of the electron. Ionization energies also exist for the higher orbits than the neutral state where an electron is removed from a negative ion.

Ionization energy and the related gain in the gravitational self energy of the nucleus appear to be same for any given electron.

When an atom gains an electron, depending upon its orbital radius, the nucleus will undergo a sudden stress in the form of expansion and as a result releases the difference in the gravitational self energy as the energy. The energy released by a gaseous neutral atom by gaining an electron is called electron affinity. Electron affinity will decrease when an electron is captured in higher orbits than its neutral state. In other words if we add more and more electrons to the negative ion, less energy will be released. Electron affinity also exists for the lower orbits of an atom than its neutral state where an electron is added to a positive ion.

Ionization energy and the electron affinity for an orbit of an atom will be same, one is gained by the atom and the other is being released by the atom. For the same orbital level, the ionization and the electron affinity will be different for different atoms.

As it has been observed earlier, chemical reactions are loosing and gaining or sharing of electrons. When an atom looses an electron and another atom gains that electron in a chemical reaction with the release of some amount of energy, how the energy equations will work?

The atom which releases the electron gains the energy and the atom which gains the electron releases the energy. These two energy entities are not equal in most of the cases. If an atom releases an electron by gaining some amount of energy then the same electron might release more energy when absorbed by another atom. Energy gained or the energy released by an atom is not dependent on the electron being transferred from one atom to another. Therefore we can easily say that the electron is not the medium or the storage of energy involved in the chemical reactions.

These energy levels were erroneously attributed to the electron shells. Electron shell containing energy is defining like an empty space containing energy. Consider a positive hydrogen ion, a single proton with empty space around it. It has no electron shells around it. When it captures an electron, the process releases the energy. According to the electron shell energy level concept, it appears that the empty space around the proton generated the energy.

The energy gained or released in chemical reactions is part of the gravitational self energy of the nucleus of the atoms involved in the reactions.

Ions are not just another simple form of the atoms. They are the small storage units for the energy. Positive ion gains the energy in the form of gravitational self energy of the compressed nucleus when it releases an electron. Negative ion looses the energy by expanding its nucleus, ultimately losing its gravitational self energy when it gains an electron. Therefore the size of the nucleus will change when an atom gains or releases an electron.

When hydrogen and oxygen atoms are combined to form the water, oxygen atom releases the energy by gaining or sharing the electrons from or with the two hydrogen atoms. Even when the carbon is reacted with the oxygen in combustion, oxygen atom only releases the energy not the carbon atom as it is widely believed. Here, two oxygen atoms are involved in the reaction therefore it releases more energy when CO_2 is formed by the combustion of the carbon. If the carbon combustion releases just the CO then the same carbon atom releases less amount of energy than when it formed as CO_2 because it has a single oxygen atom.

Gravity From A New Angle

In combustion, the amount of energy released depends upon the number of oxygen atoms involved in the reaction.

When the energy required to release the electron is less than the energy released by gaining the electron, the reaction will release a net energy. If the energy required to release the electron is more than the energy released by gaining the electron, then the reaction will consume the energy. When hydrogen and oxygen combined to form the water molecule, the reaction releases more energy than it consumes and when the water was split to form the hydrogen and oxygen, the reaction consumes the energy.

Initiation of chemical reaction

To initiate a chemical reaction that releases more energy than it consumes, it only requires releasing a single electron from an element like a metallic element which has less ionization energy. This released electron combines with the element like the oxygen and releases more energy that triggers the chain reaction in ideal conditions.

We can release the single electron by supplying any source of energy to the metallic element, like using the light, sound, shock or friction.

Change in mass in chemical reactions

Moreover, the weight of the combined elements will change in chemical reactions because of the collapse and expansion of the nucleus of the atoms involved.

When the reaction releases energy, the combined product will weigh less than the weight of the initial elements. When it consumes energy, the weight of the final product will be more than the combined weight of the initial elements. $2H_2$ and one O_2 will be heavier than the $2H_2O$ because the process releases the energy. Even though there is contraction and expansion in different atoms involved in the reactions, the amount of expansion in one atom will be more than the amount of contraction in other atom, a net loss of weight. When the hydrogen and oxygen were combined to form the water, the contraction in the common orbit of hydrogen nuclei will be less compared to the expansion of the oxygen nucleus.

The phenomenon of the change in mass or the weight in chemical reactions has already been observed in many of the earlier experiments [21]. These results were broadly disregarded because this is in conflict with another part of the standard theory called the law of conservation of mass in chemical reactions.

What is Fuel?

As explained earlier, when the carbon was burned, the energy released in the process comes from the oxygen atom not from the carbon atom. The amount of energy released depends upon how many oxygen atoms are involved in the reaction. If we call the fuel as the one which releases the energy, then in the reaction of the carbon and the oxygen to form the CO_2, which one will be called as the fuel, carbon or oxygen?

Oxygen is the real fuel which gives energy in chemical reactions. The same oxygen is the energy source for all the living things as well. The elements which have more amount of electron affinity or more electronegativity are the real fuel which generates the energy when combined with other elements.

All hydrocarbons which we all think are the fuel are just a chemical compounds. There is no importance to these hydrocarbons without the presence of the element oxygen.

The energy in the initial ignition will be used to free few electrons from the carbon or any metallic atoms. When these free electrons are combined with the oxygen atoms, more energy will be generated than used in the initial ignition. The released energy will free more electrons from the carbon atoms and eventually they combine with more atoms of the oxygen by releasing even more energy. The chain reaction will continue until all the atoms of the carbon combine with the oxygen atoms.

Gravity From A New Angle

On the earth, small ignition is sufficient to consume all the hydrocarbon compound kept in a container in seconds. But still, huge quantity of methane is surviving on Titan, one of the satellites of the Saturn, for a period of more than four billion years. Even after the heat generated by the falling space debris, the methane remained on the satellite because the element oxygen is no where near its surface.

Chapter 25:

What is Electricity?

Electricity, like the chemical reactions, involves the movement or exchange of the electrons from one place to another. As we have seen in the chapter on chemical reactions, electrons neither carry nor store the energy, then how the electricity, which was described as the flow of electrons, generates energy?

For example, lets consider a chemical battery as shown in Fig.1.

Anode is depleted in electrons and the cathode has excess electrons. As we learned earlier, positive ion will release energy when it acquires an electron.

Due to the difference in the electrical charge across the terminals, when they are connected through a conductor, electrons will flow from cathode to anode. Each electron, when it is combined with a positive ion at the anode, the ion will release energy. The energy will travel along the wire and is consumed by the resistance of the wire or will reach the cathode if there is no resistance. Energy

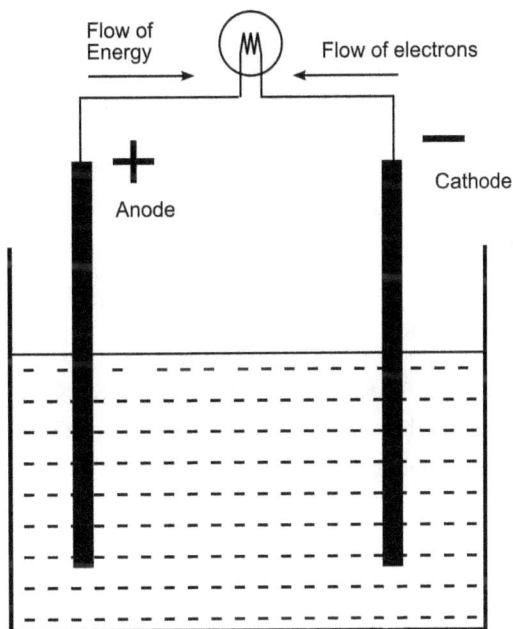

Figure 1: A chemical battery

generates at the anode until all the positive ions acquire the electrons to become neutral.

If we define the electricity as the flow of electrons, sure the electrons do travel from cathode to anode. And if we define the electricity as the flow of energy, then it travels from anode to cathode. Yes it does travel from positive to negative like all other scenarios in the nature.

Radius of the atom will decrease when it is positively charged, eventually increasing the weight of the atom. If we take out more and more electrons, atom or the positive ion becomes heavier. As we take more and more electrons away from the object, the gravity between each of the atoms of the object will increase. The shape of any object is the strength between its atoms. If the force between the atoms of an object increases, the shape of the object will crumble into a pile.

In this state, the object acts like a storage unit for the energy, like a battery or cell. When electrons are supplied to the positively charged object, it releases the energy and the atoms inside the object regain their normal state.

Similarly, if we add excess number of electrons to an atom, it releases energy and at the same time its nucleus will expand, ultimately the atom becomes lighter. As more and more electrons are added to the atom, it releases energy and becomes even lighter.

To make the object normal, we need to take out the extra electrons from the object. It requires energy to remove these extra electrons from the negative ion.

In comparison, energy required to remove electrons is higher than the energy released by adding electrons to a neutral atom of an element. If we move electrons from one object to another object made of same material then the two object system stores net energy because of the difference in the energy required for releasing and gaining the electrons by these two charged objects. If we assume that these two objects are the two plates in a capacitor then the capacitor stores more energy than it releases while charging. The same capacitor releases more energy than it consumes while it is discharging.

Weight of the anode will increase when the capacitor is charged because it is filled with positive ions. And at the same time the weight of the cathode will reduce because of the negative ions in that plate. The change in the distribution of weight in the capacitor might ultimately affect the net weight of the capacitor [22,23].

How the electrical energy propagates?

Energy generated by the anode by capturing the adjacent electron will be transferred to the adjacent atom. Subsequent capture of the electron by this new atom from its adjacent atom makes the energy to flow toward the cathode. When the energy inside multiple atoms were transferred to less number of atoms, like from thick wire to thin wire similar to a filament in a bulb, the additional energy will get radiated from the atom and a bit reaches the cathode to release more electrons to flow towards the anode.

Chapter 26:

Iron and Nickel: Tightly Packed Nuclei

An isotope of an element is an atom of that element that has more or less number of neutrons in its nucleus than the normal, stable atom of that element. It is evident that the nuclides of iron and nickel isotopes have more binding energy per nucleon than any other element.

For a given number of nucleons (protons and neutrons) the isotope of these elements will have a very compact and tightly bound nucleus. It means, the nucleus of that isotope will have more binding energy than any other nucleus of same number of nucleons.

If a nucleus has more binding energy then that nucleus will have less mass deficit, the difference between total mass of the nucleons and the mass of the nucleus when the mass is measured as the comparative gravity. Therefore, we can simply say that the mass of the nucleus is proportional to its size. As the nucleus compresses, it measures more in mass.

Mass deficit is not limited to only the nucleus. If the object compresses, it will have less mass deficit, the difference between the total mass of all the particles in the object and the mass of the object. As the mass deficit decreases, the object measures more in mass.

Chemical and the physical properties of an element depend upon the number of basic particles and how they are placed inside the atom. Then what kind of configuration of these basic particles leads to the more compact nucleus of iron and nickel isotopes?

More than one proton can't survive in a nucleus without a binding neutron between them. If only protons are present in the nucleus, repulsion will be more between them and the nucleus itself will not form.

Because of the repulsion between the protons, they can't be part of the center of the nucleus. More over, just like the gravity, if all the positive material concentrates at the center of the atom, the strength of the attraction of the point size pack of the positively charged protons will be even more on the negatively charged electrons outside the nucleus.

Structure of the Atom

The structure of the atom might be similar to the structure of the sun, a neutron core, a layer of positive plasma and then the lighter elements such as planets in the higher orbit. Neutrons might be packed together at the center of the nucleus and the protons might be orbiting around it in shells much like the electron shells around the nucleus as shown in Fig 1. The number of protons in each shell might be similar to the electron shells. The shell structure of protons to survive, there has to be sufficient number of neutrons at the center of the nucleus to exert the gravitational force.

In hydrogen atom, the proton and the electron might be orbiting around a common center.

In each of the electron and the proton shells, as we add next pair of electron and proton in those shells, the distance between the nucleus and the added pair remains same as the other electrons or protons in that shell. As the shell is being completed, the atom progresses in binding energy. An atom of an element with complete shells will have more binding energy.

As the next pair of electron and proton added to the atom to form next element, they will form into the next higher shell. As a result the size of the atom will increase. But the additional neutron at the core exerts more strength in the nucleus.

When protons and neutrons (either individual particles or as atoms of lighter elements like lithium) are pushed together, they occupy less space. Let's assume that they were compressed to a size of a point mass. Due to the proton repulsion, it expands and at the same time gravity pulls them back. Protons and the neutrons will expand until there is equilibrium between them.

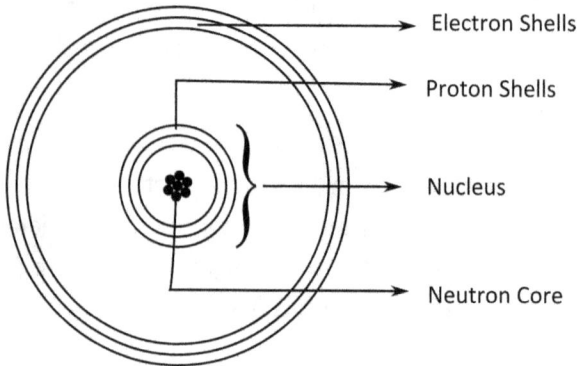

Figure 1: Structure of the atom

In the process, energy is released because the gravitational self energy of the final atom is less than the gravitational self energy of the compressed point matter of particles.

But energy is required to push the protons and neutrons into a point size object. If energy required to push them to a point is more than the energy released, total system consumes energy and the atom that formed will weigh more.

If the energy consumed is less than the energy released, the total system will release energy and the atom that formed in this process will be a lighter one.

When we compress lighter nuclei to form more compact nucleus like that of iron, more energy is consumed than it releases.

Free Energy

It's true that the energy can't be created or destroyed. It can only be transferred from one object to another or from one form to another form. Then, is there any thing like free energy in the universe?

Wherever there is an object revolves around another object, energy is involved in the system. Central object in the system continuously releases the energy to keep the peripheral object in the orbit. The energy dissipated from the sun is coming from the gravitational self energy of the core. It is being continuously depleted in energy from the time it formed.

Like the core of the solar system, neutron core at the center of the nucleus of an atom should also release energy while holding the electrons and protons in orbit around it. Gradual release of the energy will weakens the neutron core of the atom. The energy released will be more for the elements with compact nuclei like the iron and nickel.

If there are lot of electrons and protons, like the element Uranium, the loss of energy might be even more. The depletion in binding energy as the nucleus expands might be resulting in the disintegration of the atom in radioactive decay.

Because the atoms are continuously releasing the energy, we are unable to cool the objects to an absolute zero temperature. As the atomic number increases, this might become more difficult to reach the absolute zero.

Energy appears to be free only until we know the source of the energy from where it is being depleted. As the perpetual motion is depleting the energy from the earth and the light from the sun is depleting the energy in its core, the energy from an atom depletes the energy within its core. Further studies are required to find the usability of this energy.

Chapter 27:

Fission and Fusion: Which Releases the Energy?

Among the two nuclear reactions fission and the fusion, only the fission or splitting an atom was confirmed as source of energy. Fusion or combining two atoms into a single atom was never proved to release more energy than it consumes to make that happen. Will the fusion ever release energy more than it consumes?

Fission of an element releases energy because a tightly bound nucleus was split apart into lighter elements. The difference between the internal gravitational self energy of the parent element and the total internal gravitational self energy of the pack of daughter elements is released as the energy.

In the case of fusion, individual atoms were pressed together to form a new element. The internal gravitational self energy of the pack of fusion elements (individual atoms together as a pack, it is the combined internal gravitational self energy of the pack) is less than the internal gravitational self energy of the final element, therefore the process can only consume energy, it wouldn't release any energy at all. Energy will be released only when there is expansion in the substance, like the denser water turning into lighter ice crystal. In the fusion, the pack of individual elements compressed together to form a new denser element. This will never release energy.

Fusion will also release energy when the fusion elements are packed together within a density more than the density of the final fused element. Plasma, a pack of protons or the packed positive hydrogen ions is denser than the nucleus of the element lithium. When the plasma and the neutrons are combined in a fusion, the difference in the gravitational self energy of the plasma and the lithium nucleus will be released as the energy. So far in all the experiments, the energy consumed in the process of creating the plasma is more than the energy released as part of the fusion process.

Apart from plasma, if any other elements are fused together to form a new element, the process will only takes energy. For example, when the hydrogen isotopes, deuterium and tritium are fused together to form the lithium atom, the process takes more energy than it releases.

The cold fusion, the notion of combining the lighter elements at room temperature to release the energy will remain cold forever. All the earlier reports of achieving the table top fusion of lighter elements might be true. These experiments might have consumed or released a little amount of energy and it is natural.

Chapter 28:

Water, Ice and the Energy in-between

Water is a liquid form of the compound H_2O. Liquid form is a densely packed free moving molecules or atoms of a substance. Ice crystal is a hardened form of the chemical H_2O. In a crystal, each molecule or atom maintains a distance from another molecule. Crystalline form of any object occupies more space or has more volume than the liquid form of the same substance, like ice crystal occupying more space than the same amount of liquid water.

Now we know that any object with less volume has more internal gravitational self energy than the object which has more volume with same amount of material. When we cool the water, energy from the water molecules was removed and slowly they come to rest within the container. When molecules are at rest they tend to align with other molecules where the gravity is more.

Gravity changes on the surface of a molecule depending upon its shape. Pointed surfaces like a tip of a cone will have more gravity than the side of a cone. When molecules are re-aligned in equilibrium, distance between them will increase, ultimately decreasing the internal gravitational self energy. The difference in the internal energy to the water and the ice crystal will be released in the process.

If the water is contained in an object, the energy released in the process of crystallization will break the object. If the water is in open field like in a pond, the released energy evaporates some of the water. The evaporation we see on the surface of a lake in a winter in cold places like Minnesota is the result of the energy released in crystallization of the water.

In lakes, when water turned into ice, volume might not increase much because lot of water escaped from the lake as evaporation in the process of cooling the water.

An interesting reasoning for this scenario of water becoming ice is how the energy is released when the energy itself is being taken out of the system [5]. The energy released by the expanding ice will break the container in which the liquid was stored.

Every object will have a gravitational self energy which is the gravity between each and every molecules of the object to another molecule. And this energy depends upon the size of the object. Water also will have its own gravitational self energy depending upon its volume. One obvious fact when water turned into ice is the increase in the distance between each molecule to another. Object released the energy when the distance increased between the molecules. When we supply energy to the ice by heating it, the energy will be used to break the bond between the molecules of the ice crystal. When the ice collapses into the water, the object gains more gravitational self energy and stores it in the form of reduced distance between molecules.

When we measure the weight of the crystal ice with an equal amount of water from which it turned into the ice, the crystal ice will weigh less. The difference might be very small when we convert small amount of water into ice, but it is obvious. This phenomenon itself is a strong case for invalidating the shell theorem. Then how we missed to capture the difference in weight between ice and water?

Instead of precisely measuring the weight of the ice, we attempted at measuring the density of the ice because we thought we know the mass of the ice from the water where it came from. Ice is certainly less dense than the water because it floats on the water. If we increase the volume of an object, it will become less dense. We thought the same thing is happening in this case of water turning into ice. The problem appeared to be solved. But the energy involved in the process was never accounted for.

The way we measure the density of the object, volume divided by mass is not correct. Different objects of same material with same amount of volume will measure differently for the mass depending upon their shape. The calculation for the density of these objects differs depending upon the shape even when all the objects were made of same material.

Archimedes principle states that the weight of a floating object is equal to the weight of the water displaced by the same object. This principle will be true only when the weight of the displaced water is measured in the same shape and orientation as displaced by the floating object. A floating vertical bar displaces the water in a shape of a vertical bar. A floating sphere displaces the water in a shape of half sphere.

If a liquid with a freely moving molecules or atoms, like in the molten iron or liquid water, is cooled all of a sudden then it will release its internal energy at once and expand into a crystalline structure.

Liquid form is heavier and soft because the molecules were not bound to any other molecule. These molecules have kinetic energy associated with them. If we apply a small amount of energy to a molecule in a liquid, the molecules slide fast each other. Crystalline form of an object is lighter and stronger. They are hard and brittle too. If enough force is applied to the object it will break but will never bend in the process.

Chapter 29:

Formation of a Crystal

As discussed earlier, gravity between the atoms of a substance will be more when it is in liquid form because the distance between each atom is less. When the same object turns into crystalline form, the distance between the atoms will increase and will have less gravity between them. Still the crystal will be stronger than the liquid of the same substance.

All the atoms in a liquid will be continuously sliding fast each other. In a crystal, all the atoms are locked in with each other in equilibrium. More over when the atoms in a liquid distanced themselves from each other to form the crystal, the loss in the internal gravitational self energy will be released as energy. In essence, the crystal will have less gravitational self energy but will be stronger than its liquid counterpart.

Once the small crystal cell is formed, depending on its structure and the shape, the other cells will attach to it where there is more gravity, like at the corner or tip of the crystal cell.

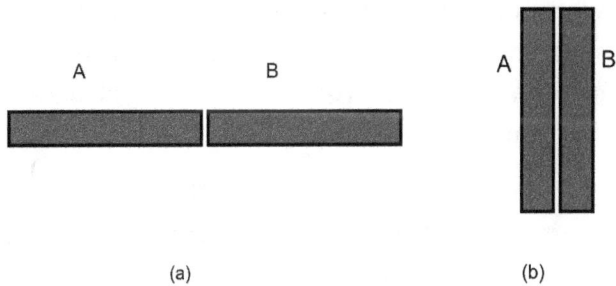

Figure 1: Strength between two crystal cells

When two lengthy bars equal in all measures are connected edge to edge as shown in Fig.1 (a), the gravity between those two bars is stronger than when they are connected side by side as shown in the Fig. 1(b).

This is the reason why we can't be able to make any scratch on the crystal structure horizontal to its base when the cells are connected to base to base. We could able to make the scratch on the sides of the crystal because the strength holding the crystal cells is weak at that point.

It is also not easy to pass the electricity through the crystal because the shape and structure of each of the atoms in the cell is in equilibrium.

Addition or removal of electrons from an atom in the crystal cell will change the strength of gravity between the atom and each of the other atoms in the structure. As a result the equilibrium will be lost between the atoms and ultimately the crystalline structure will collapse into liquid form.

Crystal forms when a liquid is suddenly expanded by releasing the gravitational self energy. Even though the crystal has less gravitational self energy than its liquid counterpart, it is stronger than the liquid because all the atoms are tightly bound to each other in equilibrium.

When a liquid object with high kinetic energy among its molecules is suddenly cooled, it will align on the lines where the strength of gravity is more. When it reaches equilibrium among all the atoms, it will remain in that form as a crystal.

Every liquid will form into crystalline structure when it releases the internal gravitational self energy. Only thing required is that the atoms should be able to form a stable electron bonds with neighboring atoms or should have the complete electron shells like the inert gases.

When electrons are continuously changes between atoms, their strength with other atoms will also change dynamically. It is difficult to attain equilibrium in this scenario.

Is solid a state? ... Is crystal a solid?

What is the state of iron and glass at the room temperature? Immediate answer for this question is solid state.

Karunakar Marasakatla

Any substance in the nature is said to have in three different states depending on energy associated with the object. Those are gaseous, liquid and solid states. Gas is a state where the individual atoms are set free because of the high energy associated with them. As the energy in the atoms dissipates the atoms starts to coalesce into the liquid form.

When the object cooled further, it hardens and forms as a solid. When a liquid iron solidifies, all the atoms in the object remains at the same place. The difference between a solid form and the liquid form of the object is simply the hardness at a specified temperature, in this example, at the room temperature.

As such there is not much difference between a solid state and a liquid state of an iron bar. In fact they both describe same state of the object, a compact and packed state of the material. Atoms are close to each other and they slide fast each other at times. Atoms are certainly not in an equilibrium state. Hardness of the iron bar comes from the packed nature of the atoms whereas the hardness of the crystal comes from the equilibrium between all the forces of the atoms in the crystal.

Crystal is a state of the material where each atom is tightly bound to each of the neighboring atoms and occupies more space than the liquid form of same amount of atoms. It also has less gravitational self energy compared to its liquid form with the same number of atoms.

Liquid form is a compact state of a matter where the atoms slide fast each other in a free moving nature. It occupies less space than the crystal of same number of atoms; it also means that it has more internal gravitational self energy than the crystal form of the object.

Gaseous state of the object has less gravitational self energy because it occupies more volume than the liquid form.

With these observations, it appears that the state we call as "solid" is not at all an independent state. It is simply a liquid state at different temperature.

Glass, iron and gold are still in liquid form at the room temperature. If they cooled further, they turn towards crystalline structure and the size of the object will also increase. They slowly become brittle and eventually will break with the released internal gravitational self energy or it may acquire a crystalline structure.

There are still three different states for a substance; they are gas, liquid and the crystal states. Crystal form of the object will be stronger than any other form of the object. Liquid form of the object will have more gravitational self energy. Gaseous state of the object will have more kinetic energy for the atoms.

Any material can be transformed into crystal. Solid iron at room temperature is not a crystal; it is still in a liquid form. If we ever achieve the crystal form of the iron, it will be stronger than the solid iron we see at room temperature. If a cup of liquid iron is cooled quickly, it not only releases the energy it acquired while melting from an iron bar but also releases the internal gravitational self energy while expanding or attaining the crystalline structure.

Chapter 30:

Capillary Action

Capillary action is not an unknown phenomenon or a different force than what we have been discussing so far. It is the result of gravity within the liquid and the gravity between the liquid and the walls of the container. Capillary action is described in the following figures using the

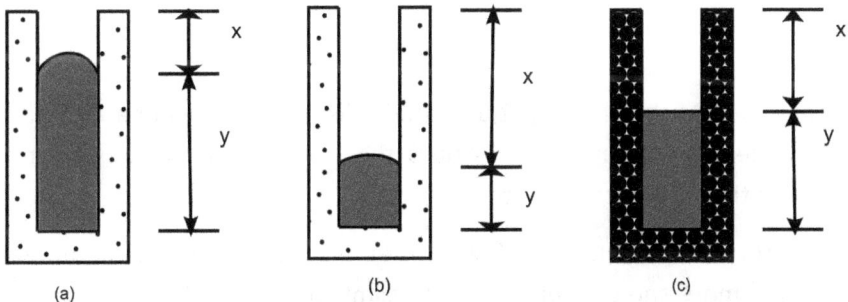

Figure 1: Capillary action in a tube

gravity between the liquid and the container it is in.

Scenario 1: As shown in Fig. 1(a), a denser liquid is contained in a lighter cylindrical container. Force within the liquid is more than the force between the liquid and the walls of the cylindrical tube. In this case, the liquid tends to form like a sphere. Because it is covered in all sides except the top, it curves upward on the open side of the tube.

Scenario 2: As shown in Fig 1 (b), if the liquid is filled to less than half of the height of the tube then the curved surface of the liquid tends to become flat due to the gravity of the walls of the entire empty part of the tube.

Scenario 3: As shown in Fig. 1 (c), if the density of the container is same as the liquid then the liquid surface turns toward a flat surface.

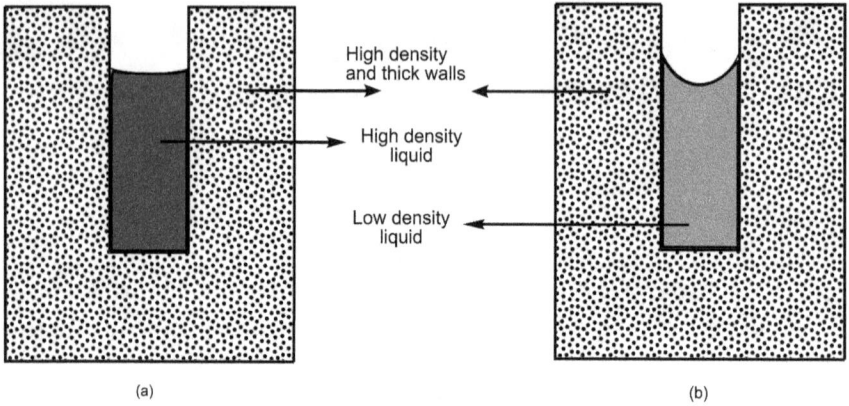

Figure 2: Capillary action in a thick wall container

Scenario 4: As shown in Fig. 2(a), if the walls of the container are thicker then the material in the walls pulls the liquid around them upward, making the surface to dip in the center.

Scenario 5: If the liquid is of low density, then the dip at the center will be even more and the liquid tends to climb upward as shown in the Fig. 2(b).

Whether the liquid curves upward or downward depends upon the thickness of the walls, density of the wall material, density of the liquid, diameter of the tube and the proportion of the filled and empty parts of the tube as shown in Fig. 3.

Gravity of the particles of the container and the entire liquid to the particles on the surface of the liquid determines the direction of the liquid particles on the surface of the liquid whether it curves upward or downward. Particle of the surface liquid moves in the direction of the resultant force of all these forces. If the empty part of the container is lengthy and the walls are thicker then the liquid climbs onto the wall. Liquid particles get their energy from the gravitational self energy of the container walls.

Every object consumes energy when that object was made, just like the energy used to melt the iron. As that object was put to use, the same energy will be dispensed when that object was under stress. Object

Karunakar Marasakatla

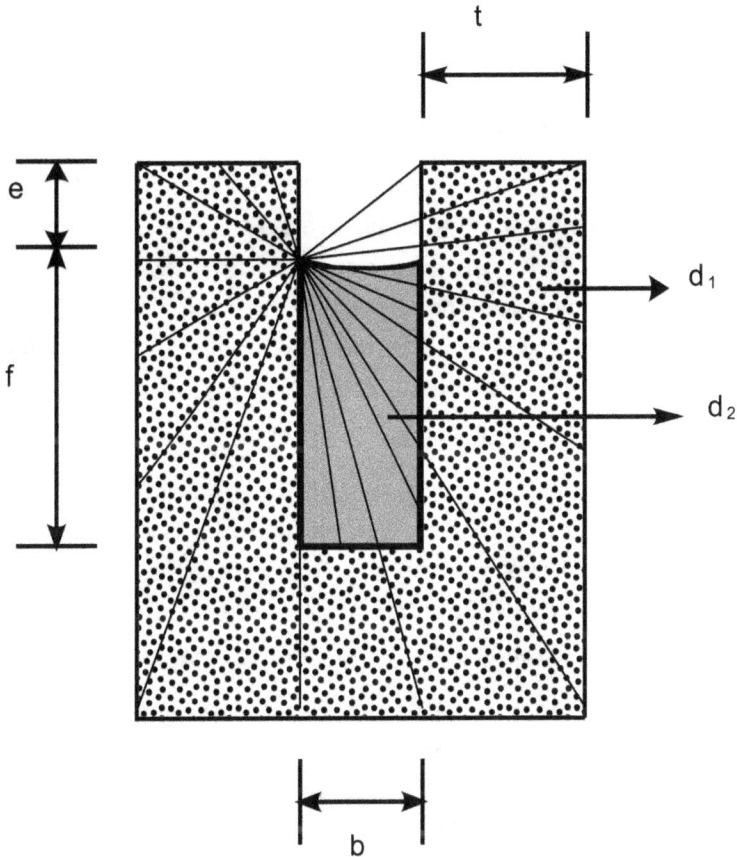

Figure 3: Capillary action and the forces involved

slowly deteriorates in strength if it is continuously used. The container uplifts the liquid as long as it has sufficient gravitational self energy.

If all the combined force points downward then the point liquid moves downward. If all the forces are equal, then the point liquid remains stationary and liquid surface remains flat.

For any given scenario, when the tube or the container is full, the liquid bends downward at the top of the tube. When the level of the liquid is reduced, at some point, the surface will become flat. Then if we reduce the liquid even more then the surface starts to dip at the center.

Gravity From A New Angle

Therefore it is not a standard for a liquid to bend or dip in the tube; it can be any of those scenarios depending upon different parameters.

What is surface tension?

Molecules on the surface are tightly bound gravitationally to each other and to the ones below them to some layers. Molecules gets added and replaced at the bottom layer of the surface. Other layers close to the surface possibly stay intact. That makes the top portion or the surface of the liquid stronger than the rest of the liquid. When we boil the liquid, water mixes up and down, this breaks the top layer, as a result the strength of the layer weakens. Surface tension is clear demonstration of how strong is the gravity within a liquid material.

Because the top layer is stationary, when the water is cooled, probably the top layer crystallizes first and then it spreads downward. Once the top layer is frozen, it blocks the release of the energy from the freezing water, so the crystallization of the remaining water stops.

Capillary action is the resultant gravity of the container and the entire liquid on the surface layer of the liquid.

Chapter 31:

Repulsive Gravity and Other Forces

Gravity sometimes even appears like a repulsive force. If a small particle O is placed very close to a thin plate P, at the center of the plate as shown in Fig 1 (a) then the differential force will be more than the displacement force. As a result the particle will fly away from the vicinity of the plate or it will be very difficult to push the particle even closure to the plate. The strong differential force appears like a repulsive force. But the same particle from a larger distance gets attracted to the same plate as shown in Fig. 1 (c).

If the surface of the plate C, is curved towards the particle O, as shown in the Fig. 2 (a) then the repulsive

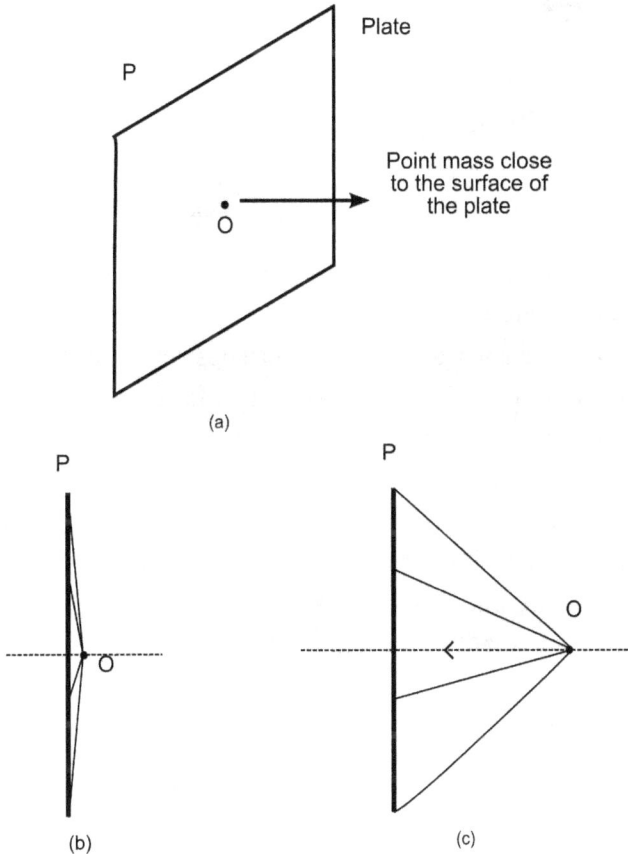

Figure 1: Force between a particle and a plate

force will be even more. In this scenario, it will be very difficult to place the particle close to the plate.

If the particle is placed at the outside of the curved plate as shown in the Fig 2 (b) then the particle gets attracted to the plate and it might even attach to the plate if the curvature of the plate is more.

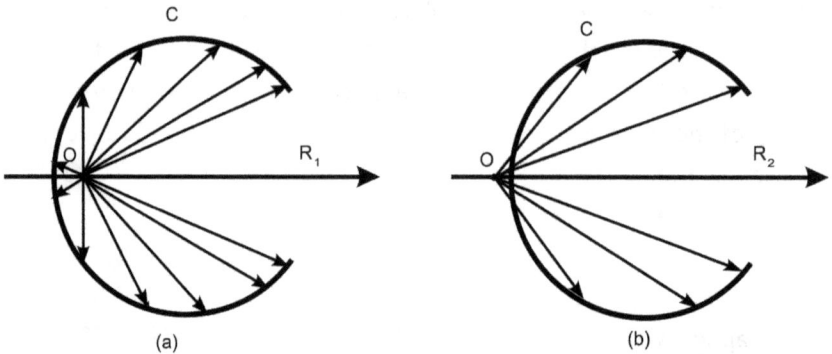

Figure 2: Gravity at a curved surface

This repulsive nature of the gravity would have been more evident if we were living on a flat earth. If we spread the whole material inside the earth in a square slab with a thickness of 1 Km, then the repulsive force at the center of the square on its surface will be very strong. If an object was kept at the center of the flat earth on its surface then that would be repulsed into the space. On the surface of the flat earth and to certain elevation, the repulsive force will be stronger than the attractive force. Within this range, it is difficult to attach any object to the surface of the flat earth. Because we all are attached to the earth, we can confidently conclude that the earth is in a shape of a sphere.

Water droplet on the Lotus leaf

Lotus leaf doesn't repel the drop of water on its surface. The self gravity of the water molecules is greater than the gravity between the water and the surface of the leaf. In this scenario, water tends to be together because there is nothing else attracts it more than itself to break the sphere of water. If we pour water on the surface of a light, thin and smooth sheet then the water might form as sphere as well.

If we pour a liquid with very low density than the water on the leaf, it might spread along the leaf.

Nanotechnology

The nano world is a strange field. All the observations suggest that there is a strong force between nano particles. Still, the theory says that there is no strong force acting at that level between these nano particles. According to the standard theory, the strong nuclear force and the gravity are out of reach at the level where these nano particles are interacting with each other.

Even at the granular level, strong cohesion was observed between two granular objects [24]. Alternating views were expressed as the cause of this cohesion between two grains.

The strong attractive or repulsive force observed at the nano or granular level can be attributed to the gravity itself and it can be explained by using this new theory of gravity.

Fifth Force

Gravity between a floating object and the earth will be nullified like a floating object in an orbit. In this scenario, the gravity between two floating objects will be clearly visible. The two floating objects on the water will naturally gets attracted to each other gravitationally. The force of the attraction depends upon the size and orientation of the objects towards each other. Because none of the existing principles explains this behavior of objects, a new fifth force was proposed as the cause of this action. The force observed between the floating objects is simply the gravity itself.

Dark Energy

Gravity exerted by the objects like the stars and the galactic center is continuously decreasing due to the energy released from these objects. Everything that orbits around these objects will gradually expand their orbits due to the loss of strength in the central object.

In the same way the orbits of the stars will also expand in the galaxy. As the galaxies expand, the strength between them also will be reduced. As a result, the galaxies also distance themselves from each other.

Is the speed at which these objects are expanding is matching with the loss of energy in the universe? This needs to be studied further.

Anti-Gravity

There is no force like the anti-gravity in the nature. Gravity can only pull the objects towards each other. It can't make the objects repulse each other.

As we have seen earlier in the chapter on electricity, weight of the object can be altered with the supply of charge to an object. Lot of negative charge makes the object lighter and lot of positive charge makes the object heavier in weight.

The only possibility to lift an object using the gravitational force is to supply the negative charge to the object. If the object becomes lighter than the air then the surrounding air will push the object upward.

Conclusion

Is this new theory for real? Is the solution to most intriguing problem so simple? Are we using the flawed basic principles in physics all along? These might be some of the common questions arise after reading this book.

For a theory to be successful, it has to explain many of the observed phenomena. There are possibly thousands of facts observed so far in science. This theory successfully explained most common observations as well as most intriguing facts.

The simplicity of the theory makes the extraordinary concepts like curvature of space, change in acceleration, push gravity or even ever expanding objects [5] irrelevant in explaining the nature of gravity. Gravity is a force of attraction as envisioned by the Newton, only exception being that it is stronger than initially assumed.

The main issues in delaying the understanding of gravity are the assumption of gravity as not a source of energy, concept of force, work and the definition of mass. All the flawed concepts derailed the progress of science.

In the period when the standard theories reined the throne, most of the observations which were not explainable by these theories were rejected as baseless based on the evaluation of the standard theories. In view of this new theory, all of the earlier observations need to be re-evaluated.

At the same time, all the established theories need to be re-evaluated too. Most of the current theories depend upon the current definition of mass, a comparative gravity, which turned out to be flawed. In this new light, all of the existing theories need to be re-evaluated.

If this new theory turned out to be true then the scope of its application will be enormous. Let's hope that this new understanding will be used to build the new environment for the betterment of humanity.

Gravity From A New Angle

References

1. G. Schilling, "Battleground Galactica: Dark Matter vs. MOND", Sky & Telescope, pp 30-36, (April, 2007).
2. M. Chalmers, "Gravity's dark side", Physics World, (January 1, 2006).
3. T. Folger, "Nailing down Gravity", Discover, 24(10), pp 34-41, (October 2003).
4. D. Rubin, Letters: "MOND matters", Discover, (October 2006).
5. M. McCutcheon, "The Final Theory: Rethinking Our Scientific Legacy", Universal Publishers, USA, (2002).
6. DOE/Sandia National Laboratories, "Kilogram is Losing Weight: Redefine Kilogram based on Universal Constants, Scientists Urge", Available at "http://www.sciencedaily.com/releases/2008/02/080228120943.htm", (28 February, 2008).
7. I. McCausland, "Anomalies in the History of Relativity", Journal of Scientific Exploration, Vol. 13, No. 2, pp 271-290, (1999).
8. M. Milgram, Astrophys. J., 270, 365 (1983).
9. M. Milgram, "Dark Matter Really exists?", Scientific American, pp 42-52, (August 2002).
10. O. Manuel, "The structure of the Solar Core", Proceedings of the Fourth International Conference on Beyond Standard Model Physics, BEYOND 2003, (2003).
11. S. Weinberg, "A Unified Physics by 2050?", Scientific American, pp 68-75, (December 1999).
12. G. Gillies, "The newtonian gravitational constant: recent measurements and related studies", Rep. Prog. Phys., 60(2), pp151-225, (February, 1997).
13. D. Kestenbaum, "The Legend of Big G", New Scientist, pp 39-42, (17 January, 1998).

14. V. Rubin, "Rotation of the Andromeda Nebula from a Spectroscopic Survey of Emission Regions", Astrophysical Journal, pp 159:379, (1970).

15. J. Anderson et al., "Indication, from Pioneer 10/11, Galileo, and Ulysses Data, of an Apparent Anomalous, Weak, Long-Range Acceleration", Physical Review Letters, 81: pp 2858-2861,(1998).

16. G. Darwin, Phil. Trans. R. Soc., 11 (170), pp447-538, (1879).

17. T. Flandern, "Dark Matter, Missing Planets and New Comets", North Atlantic Books, 1999.

18. J. Lissauer, "It's not easy to make the Moon", Nature, pp 327-328, 389, (25 September, 1997).

19. D. C. Mishra and M.B.S. Vyaghreswara Rao, "Temporal variation in gravity field during solar eclipse on 24 October 1995", Current Science, 72(11), pp 782-783 (1997).

20. M. Allais, Aero/Space Engineering, (18 (9) and (10)), (September and October, 1959).

21. K. Vokamer et al., "Experimental Re-examination of the law of conservation of mass in chemical reactions", Journal of Scientific Exploration, Vol. 8, No. 2, pp 217-250, (1994).

22. C. Platt, "Breaking the law of Gravity", Wired, 6.03, (March, 1998).

23. J. Woodward, Foundations of Physics Letters, 3(5), (1990).

24. Amarouchene et al., "Capillay-like Fluctuations at the Interface of Falling Granular Jets", Physical Review Letters, 100(21), (2008).

Change History

Version: 1.0
Changes: Initial version.
Published: March, 2009 (eBook)

Version: 1.1
Changes: All the chapters in Part-I were restructured. Some of the chapters in Part-II were moved into the Part-I to further explain the flaws in the current theories. Shell theorem was analyzed in detail with a diagram. Formation of the moon also revised with additional diagram and a new reference.
Published: July, 2009 (eBook)

Version: 1.2
Changes: There are minor changes in this version. The entire text has been reformatted for printing.
Published: August, 2009 (eBook, paperback)

Index

Gravity From A New Angle

www.ingramcontent.com/pod-product-compliance
Lightning Source LLC
Chambersburg PA
CBHW060030210326
41520CB00009B/1076